Plasma Chemistry and Gas Conversion

Edited by Nikolay Britun and Tiago Silva

Published in London, United Kingdom

IntechOpen

Supporting open minds since 2005

Plasma Chemistry and Gas Conversion
http://dx.doi.org/10.5772/intechopen.76273
Edited by Nikolay Britun and Tiago Silva

Contributors
Guoxing Chen, Annemie Bogaerts, Ramses Snoeckx, Georgi Trenchev, Weizong Wang, Tomohiro Nozaki,
Zunrong Sheng, Seigo Kameshima, Kenta Sakata, Nikolay Britun, Tiago Ponte Silva

Notice
Statements and opinions expressed in the chapters are these of the individual contributors and not
necessarily those of the editors or publisher. No responsibility is accepted for the accuracy of
information contained in the published chapters. The publisher assumes no responsibility for any
damage or injury to persons or property arising out of the use of any materials, instructions, methods
or ideas contained in the book.

First published in London, United Kingdom, 2018 by IntechOpen
IntechOpen is the global imprint of INTECHOPEN LIMITED, registered in England and Wales,
registration number: 11086078, The Shard, 25th floor, 32 London Bridge Street
London, SE19SG – United Kingdom
Printed in Croatia

British Library Cataloguing-in-Publication Data
A catalogue record for this book is available from the British Library

Additional hard copies can be obtained from orders@intechopen.com

Plasma Chemistry and Gas Conversion
Edited by Nikolay Britun and Tiago Silva
p. cm.
Print ISBN 978-1-78984-840-3
Online ISBN 978-1-78984-841-0

We are IntechOpen,
the world's leading publisher of
Open Access books
Built by scientists, for scientists

3,900+
Open access books available

116,000+
International authors and editors

120M+
Downloads

151
Countries delivered to

Our authors are among the

Top 1%
most cited scientists

12.2%
Contributors from top 500 universities

Interested in publishing with us?
Contact book.department@intechopen.com

Numbers displayed above are based on latest data collected.
For more information visit www.intechopen.com

Meet the editors

Nikolay Britun graduated from Kiev National University, Ukraine, in 2002 and received a PhD degree from Sungkyunkwan Univeristy, South Korea, in 2008. He is currently working in the laboratory "Chimie des Interactions Plasma-Surface" at the University of Mons, Belgium. His research interests are related to plasma spectroscopy, plasma chemistry, and in particular to diagnostics of the processes related to CO_2 decomposition in non-equilibrium discharges.

Tiago Silva earned his master's degree in Engineering Physics at IST in 2012 and his PhD degree at the University of Mons, Belgium, in 2015. He currently holds a postdoctoral position at IPFN/IST. His research interests are related to the diagnostics and modeling of plasma sources in view of their optimization and plasma-based conversion of greenhouse gases into valuable chemicals.

Contents

Chapter 1

Introductory Chapter: Plasma Chemistry for Better CO_2 Conversion

Nikolay Britun and Tiago Silva

1. Preface

The aim of this book is to cover recent advances in the field of plasma chemistry and, in particular, to explore the role of low-temperature discharges for efficient greenhouse gas conversion, synthesis of valuable chemicals, potential fuel production, and storage.

Low-temperature plasmas produced by electric discharges represent unique non-equilibrium state of matter where electrons possess much higher temperatures than neutrals and ions. This distinctive feature opens new possibilities for production of highly reactive species in a chemically rich environment close to the room temperature in the wide range of gas pressure, which may vary from mTorr range to a fraction of atmosphere. Moreover, the discharge conditions far from thermodynamic equilibrium may further intensify the "traditional" chemical processes, which normally happen without plasma.

Low-temperature discharges are deeply related to a large number of important technologies with extraordinary societal and environmental benefits. For example, since the second half of the nineteenth century, low-temperature plasma has been used to improve the microelectronics industry [1]. Indeed, these discharges are able to provide ion fluxes that are responsible for surface modifications by sputtering, etching, activation, and deposition, which are of a critical importance for the development of any micro device. Other domains in which low-temperature plasmas play an important role involve light sources [2], lasers [3], sterilization of biological samples [4], etc. All of these technologies make important contributions to the development of the modern society.

2. Historical remarks

The use of low-temperature plasmas for chemical conversion of the greenhouse gases has a rich history and can be traced back to the 1970s–1980s, namely, to the research related to CO_2 transformation into the valuable chemicals conducted in the former USSR (see [5] and therein). During this period the experimental and theoretical background on the plasma-chemical processes has been mainly understood (see [6] and therein). Interestingly, it was already estimated theoretically that the limit of the energy efficiency of CO_2 decomposition (term defined in the following chapters) in microwave discharges can reach about 43% in the equilibrium regime and about 80% in the non-equilibrium regime [5]. The corresponding experimental results related to these efficiencies have shown excellent agreement with the estimates [5–7].

Nowadays the gamut of potential applications of plasma-chemical processes undergoes significant widening, covering in the case of CO_2 decomposition the areas from the treatment of power plants exhausts [6] to the potential fuel production on Mars [8]. Regarding the environmental concerns associated with consumption of the fossil fuels, this topic is now also receiving a special attention. Naturally, these concerns are related to the necessity of moving toward the usage of the renewable technologies that would give access to the green CO_2-based electricity. In this case, the non-equilibrium discharge can potentially act as a vehicle transforming electricity into the useful chemical reactions, being at the same time environmentally friendly. This paradigm is well recognized by the modern plasma research community due to its important social and economic footprints [1].

3. How does it work?

One particular plasma-chemical process, which is mainly covered in this book, as well as widely studied nowadays in general, is the plasma-based dissociation of CO_2 into CO and O. Such a dissociation process may be considered as a first step toward the production of fuels and chemical feedstock, for example, methanol. Under this scenario an efficient plasma-based CO_2 decomposition process can provide a suitable storage solution for renewable sources via the conversion of temporary electrical energy. This would permit fuel production using electrical power in remote locations where solar/wind energy availability is optimal, or even abundant, and to use the existing infrastructure for energy distribution to the end users [9]. From this point of view, CO_2 would no longer be considered as a pollutant but rather a raw material for further transformations using the plasma technology.

Motivated by the previously mentioned strategy, numerous research groups are currently focused on achieving the maximum conversion and energy efficiencies (defined in the following chapters) associated to the CO_2 decomposition via modeling [10, 11] as well as through the experimental studies [12–15]. In addition to the plasma itself, utilization of the pre-activated highly porous catalyst may also significantly increase the energy efficiency of the conversion process, as shown in the numerous literature sources [16, 17] and demonstrated in this book.

The main idea behind the mentioned research works is to take advantage of the non-equilibrium nature of low-temperature plasmas, with activation of the plasma at low-energy cost. Indeed, it is relatively easy to transfer energy to the CO_2 vibrational excitation using a plasma source possessing low electron temperature (~1 eV). Under this scenario, it is possible to benefit from the energy stored in the vibrational levels, that is, vibrational excitation of the molecule, which is known to be favorable for molecular decomposition in the case of CO_2 [6]. More specifically, if the electron energy is selectively channeled into the CO_2 asymmetric stretch mode of vibration, then the vibrational quanta can be pumped up through the so-called vibrational "ladder climbing" mechanism, offering a unique way to achieve efficient decomposition [10].

4. Achievements and challenges

As mentioned already, rather high-energy efficiency of CO_2 conversion had been reached in the past using the supersonic gas flow (for gas expansion and cooling) in a microwave plasma [7]. This result had a huge impact on the plasma research community, which nevertheless were not reproduced since then. Other types of plasma reactors including dielectric barrier discharges (DBD), gliding arc plasmatrons

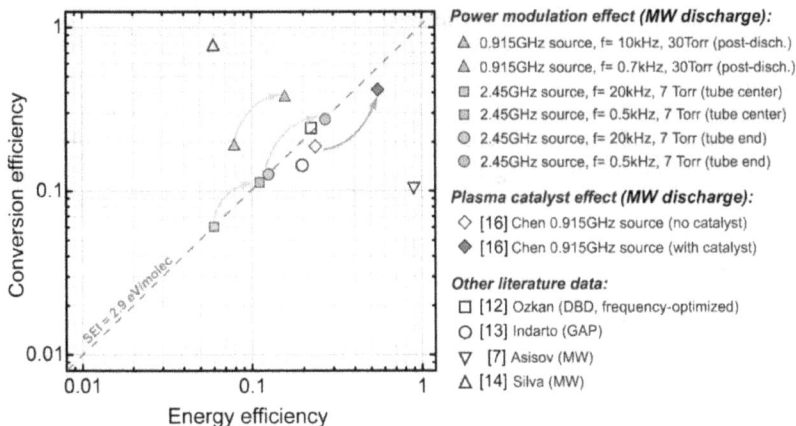

Figure 1.
The optimized values of the CO_2 conversion efficiency and energy efficiency obtained in various low-temperature discharges. The arrows correspond to the efficiency gains achieved by using power modulation (green) and plasma catalysis (red). Reproduced with permission from Ref. [17].

(GAP), and microwave plasmas have been used, together with plasma catalysis, having a goal of increasing the energy efficiency associated to the CO_2 conversion (see **Figure 1**). These studies have shown that such an increase in energy efficiency of the decomposition can be attained through the fine-tuning of different plasma parameters such as gas pressure, temperature, gas composition, molecule residence time, etc. More recent works have proven that using the plasma power interruption along with plasma catalysis also enhances the CO_2 conversion and energy efficiencies significantly, as illustrated in **Figure 1**.

Despite the relevance of the abovementioned works, the application of plasmas for large-scale fuel production is not yet viable [18]. Indeed, the issues related to the optimal plasma operation conditions (such as gas pressure, reactor geometry, degree of non-equilibrium, etc.) still provoke many questions, while the pathway of CO_2 dissociation is not yet completely understood to achieve its full control and optimization. In order to overcome these difficulties, the scientific research toward both modeling and diagnostic studies is mandatory for a deeper understanding of these processes.

From the modeling point of view, there are still many challenges ahead. Among these challenges, there is a lack of modeling studies related to calculation of the rate coefficients in which CO_2 ro-vibrational excitation is present. In this respect, many CO_2 chemical reaction rates are still poorly known. This needs further attention from the theoretical methods (e.g., based on quasi-classical trajectory simulations [19]) to calculate rate coefficients more accurately than those which are currently available in literature. These calculations are essential to simulate and predict the overall behavior of CO_2 discharges while guiding future experiments targeted at fuel production.

From the experimental point of view, careful verification of the role of electronic and vibrational states of CO and CO_2 molecules as well as the electronic states of O atoms during the CO_2 decomposition process under different degrees of discharge non-equilibrium may shade light on the valuable decomposition pathways pointing out to an optimum regime of plasma operation for maximization of the energy efficiency of decomposition and other critical parameters. In this case, the implementation of the advanced (time- and space- resolved) laser diagnostics along with the discharge characterization by infrared absorption spectroscopy would be critical.

As a conclusion we can say that, the development of the low-temperature plasma sources operating at the elevated (virtually up to atmospheric) pressure is mandatory in order to scale up the laboratory processes and to match the industrial workflow in gas conversion. Thus, the experimental research related to production of valuable fuels and chemical feedstock (e.g., hydrocarbon-based fuels) through CO_2-containing discharges is mandatory. Besides this, the other approaches enabling reasonably high-energy efficiencies of conversion should be investigated as well.

The chapters gathered in this book are dedicated to the carbon dioxide (CO_2) and methane (CH_4) decomposition in non-equilibrium plasma discharges. These chapters represent the state-of-the-art modeling and experimental research studies in the field, paying special attention to the plasma catalyst and containing recent achievements in the field of greenhouse gas conversion in non-equilibrium discharges. The book can be considered as an addition to the more general book "Green Chemical Processing and Synthesis" (INTECH, DOI: 10.5772/65562) published recently.

Author details

Nikolay Britun[1*] and Tiago Silva[2]

1 University of Mons, Mons, Belgium

2 University of Lisbon, Lisbon, Portugal

*Address all correspondence to: nikolay.britun@umons.ac.be

IntechOpen

References

[1] Adamovich I, Baalrud SD, Bogaerts A, et al. The 2017 plasma roadmap: Low temperature plasma science and technology. Journal of Physics D: Applied Physics. 2017;**50**:323001

[2] Gibson ND, Kortshagen W, Lawler JE. Investigations of the 147 nm radiative efficiency of Xe surface wave discharges. Journal of Applied Physics. 1997;**81**:1087

[3] Moutoulas C, Moisan M, Bertrand L, Hubert J, Lachambre JL, Ricard A. A high-frequency surface wave pumped He-Ne laser. Applied Physics Letters. 1985;**46**:323

[4] Kelly-Winterberg K, Montie TC, Brickman C, Roth JR, Carr AK, Sore K, et al. Room temperature sterilization of surfaces and fabrics with a one atmosphere uniform glow discharge plasma. Journal of Microbiology and Biotechnology. 1998;**20**:69

[5] Rusanov VD, Fridman AA, Sholin GV. The physics of a chemically active plasma with nonequilibrium vibrational excitation of molecules. Soviet Physics Uspekhi. 1981;**24**:447

[6] Fridman AA. Plasma Chemistry. New York: Cambridge University Press; 2005

[7] Asisov RI, Vakar AK, Jivotov VK, et al. The translation is: Nonequilibrium plasma-chemical process of CO_2 decomposition in a supersonic microwave discharge. In: Russian Proceedings of the USSR Academy of Sciences. 1983;**271**:94

[8] Guerra V, Silva T, Guaitella O. Living on mars: How to produce oxygen and fuel to get home. EPN. 2018;**49**:15

[9] Goede APH. CO_2-neutral fuels. EPJ Web of Conferences. 2015;**98**:7002

[10] Silva T, Grofulović M, Klarenaar BLM, Morillo-Candas AS, Guaitella O, Engeln R, et al. Kinetic study of low-temperature CO_2 plasmas under non-equilibrium conditions. I. Relaxation of vibrational energy. Plasma Sources Science and Technology. 2018;**27**:015019

[11] Kozák T, Bogaerts A. Splitting of CO_2 by vibrational excitation in non-equilibrium plasmas: A reaction kinetics model. Plasma Sources Science and Technology. 2014;**23**:045004

[12] Ozkan A, Dufour T, Silva T, Britun N, Snyders R, Reniers F, et al. DBD in burst mode: Solution for more efficient CO_2 conversion? Plasma Sources Science and Technology. 2016;**25**:055005

[13] Indarto A, Yang DR, Choi JW, Lee H, Song HK. Gliding arc plasma processing of CO_2 conversion. Journal of Hazardous Materials. 2007;**146**:309

[14] Silva T, Britun N, Godfroid T, Snyders R. Optical characterization of a microwave pulsed discharge used for dissociation of CO_2. Plasma Sources Science and Technology. 2014;**23**:025009

[15] Chen G, Georgieva V, Godfroid T, Snyders R, Delplancke-Ogletree MP. Plasma assisted catalytic decomposition of CO_2. Applied Catalysis B: Environmental. 2016;**190**:115

[16] Chen G, Britun N, Godfroid T, Georgieva V, Snyders R, Delplancke MP. An overview of CO_2 conversion in a microwave discharge: The role of plasma-catalysis. Journal of Physics D: Applied Physics. 2017;**50**:084001

[17] Britun N, Chen G, Silva T, Delplancke MP, Snyders R. Green Chemistry Processing and Synthesis. In: Karame I, Srour H, editors. Croatia, Rijeka: InTech; 2017. pp. 3-27

[18] van Rooiji GJ, Akse HN, Bongers WA, van de Sanden MCM. Plasma for electrification of chemical industry: a case study on CO_2 reduction. Plasma Physics and Controlled Fusion. 2018;**60**:014019

[19] Bartolomei M, Pirani F, Laganà A, Lombardi A. A full dimensional grid empowered simulation of the $CO_2 + CO_2$ processes. Journal of Computational Chemistry. 2012;**33**:1806

Chapter 2

Modeling for a Better Understanding of Plasma-Based CO_2 Conversion

Annemie Bogaerts, Ramses Snoeckx, Georgi Trenchev and Weizong Wang

Abstract

This chapter discusses modeling efforts for plasma-based CO_2 conversion, which are needed to obtain better insight in the underlying mechanisms, in order to improve this application. We will discuss two types of (complementary) modeling efforts that are most relevant, that is, (i) modeling of the detailed plasma chemistry by zero-dimensional (0D) chemical kinetic models and (ii) modeling of reactor design, by 2D or 3D fluid dynamics models. By showing some characteristic calculation results of both models, for CO_2 splitting and in combination with a H-source, and for packed bed DBD and gliding arc plasma, we can illustrate the type of information they can provide.

Keywords: CO_2, plasma chemistry, plasma reactor, fluid dynamics modeling, chemical kinetic modeling

1. Introduction

In recent years, there is increasing interest in plasma-based CO_2 conversion [1]. Several types of plasma reactors are being investigated for this purpose, including (packed bed) dielectric barrier discharges (DBDs) [2–10], microwave (MW) plasmas [11–13], and ns-pulsed [14], spark [15], and gliding arc (GA) [16–20] discharges. Research focuses on pure CO_2 splitting into CO and O_2, as well as on mixtures of CO_2 with a hydrogen source, such as CH_4 but also H_2O or H_2, to produce value-added chemicals like syngas, hydrocarbons, and oxygenated products. Key performance indicators are the conversion and the energy efficiency of the process, as well as selectivity to produce specific value-added chemicals. To realize the latter, the plasma should be combined with a catalyst (e.g., [3–5, 21]), as the plasma itself is a too reactive environment and thus not selective.

To improve the application, a good insight in the underlying mechanisms is crucial. This can be obtained by experiments, but modeling the plasma chemistry and reactor design can be a valuable alternative, as it provides information on the most important chemical reaction pathways and on how the geometry and operating conditions can be optimized to improve the results.

In this chapter, we will describe the basics of both plasma chemistry modeling (typically based on 0D chemical kinetic models) and plasma reactor modeling

IntechOpen

(typically based on 2D, or even 3D, fluid models), and we will show some characteristic examples from our own research, to illustrate how such models can give more insight in the underlying mechanisms. First, however, we will present a brief overview of the different models relevant to CO_2 conversion that have been reported in literature.

2. Literature overview on modeling for plasma-based CO_2 conversion

Describing a detailed plasma chemistry in 2D or 3D models, with 100s of species and chemical reactions, is not yet feasible, due to excessive calculation times. Therefore, a detailed plasma chemistry is typically described by 0D chemical kinetic models or sometimes by 1D fluid models. The first papers on CO_2 plasma chemistry modeling were published back in 1987–1995 but were applied to CO_2 lasers [22–24]. Some papers also studied the vibrational kinetics of CO_2 for gas flow applications [25, 26]. Rusanov et al. [27] were the first to develop a model for CO_2 conversion in a MW plasma, based on particle and energy conservation equations for the neutral species, and an analytical description of the vibrational distribution function.

In the last decade, the research on plasma-based CO_2 conversion experienced a clear revival, and quite some plasma chemistry models have been developed in literature, for either pure CO_2 splitting [7, 28–48] or CH_4 (of interest for hydrocarbon reforming) [49–52], as well as in various mixtures, that is, CO_2/CH_4 [53–66], CH_4/O_2 [66–72], CO_2/H_2O [73], and CO_2/H_2 [74, 75], of interest for producing value-added chemicals, or in mixtures of CO_2/N_2 [76, 77] or CH_4/N_2 [78–83], more closely mimicking reality, as N_2 is a major component in effluent gases. Recently, we gave an overview of such 0D models for plasma-based CO_2 and CH_4 conversion [84], and we also presented a very comprehensive plasma chemistry model for CO_2 and CH_4 conversion in mixtures with N_2, O_2, and H_2O [85]. These plasma chemistry models can provide detailed information on the underlying chemical reaction pathways for the conversion or product formation.

Furthermore, to investigate which reactor designs can lead to improved CO_2 conversion, 2D or even 3D fluid models can be used; they offer a good compromise between level of detail and calculation time. To our knowledge, the number of 2D models for describing CO_2 conversion is very limited [86, 87], and there exist no 3D models yet for this purpose. Most of the 2D/3D fluid models developed up to now in the literature for the typical plasma reactors used for CO_2 conversion are developed in argon or helium, or sometimes air, with limited chemistry, to reduce the calculation time.

For packed bed DBD reactors, different types of modeling approaches have been developed. Chang [88] presented a 0D plasma chemistry model, simply predicting the enhancement factor of the electric field in the voids between the packing pellets from the ratio of the dielectric constant of the pellets and the gas phase. Takaki et al. [89] applied a simplified time-averaged 1D model in N_2, based on solving the transport equations and Poisson's equation. Zhang et al. [90] performed 2D particle-in-cell/Monte Carlo collision (PIC/MCC) simulations for the filamentary discharge behavior in a parallel-plate packed bed DBD reactor in air. Kang et al. [91] developed a 2D fluid model for a DBD with two stacked ferroelectric beads and studied the propagation of the microdischarges, but no plasma species were explicitly considered. Russ et al. [92] applied a 2D fluid model for studying transient microdischarges in a packed bed DBD operating in dry exhaust gas. Based on a 2D fluid model for a packed bed reactor with dielectric rods, Kruszelnicki et al. [93] presented a very interesting and detailed study on the mechanism of discharge propagation in humid air, reporting that the discharges can generally be classified in three modalities: positive restrikes, filamentary microdischarges, and surface ionization waves. They observed that the type of discharge dominating the production of reactive species depends on the dielectric facilitated electric field enhancement, which is determined by the topography and orientation of the dielectric lattice. Finally, they demonstrated that photoionization plays an important role in discharge propagation through the dielectric lattice, because it seeds initial charge in regions of high electric field, which are difficult to access for electrons from the main streamer [93]. Van Laer et al. [94–96] developed two complementary 2D fluid models to describe a packed bed DBD in helium, to elucidate the electric field enhancement between the packing beads, and the effect of the dielectric constant of the packing beads, as well as the gap size and bead size. Wang et al. [97] applied a 2D fluid model to a packed bed DBD in air, studying the behavior of positive restrikes, filamentary microdischarges, and surface discharges, as well as the transition in discharge modes upon changing the dielectric constant of the packing beads. Finally, Kang et al. [98] also presented a 2D fluid model to study surface streamer propagation in a simplified packed bed reactor, in comparison with experimental data, obtained from time-resolved ICCD imaging.

For MW plasmas, a large number of models were presented in the literature, and we refer to [99] for a recent overview. Van der Mullen et al. [100–102] as well as Graves et al. [103] developed self-consistent 2D fluid models, based on Maxwell's equations for the electromagnetic field and plasma fluid equations, assuming ambipolar diffusion. Some of these models were applied to intermediate pressure coaxial microwave discharges [102], while others describe atmospheric pressure cylindrical (surfaguide or surfatron) MW plasmas [101, 103]. Although being very valuable, these models did not apply to the application of CO₂ conversion. Recently, Georgieva et al. [99] performed a comparison between two fluid models, based on the coupled solution of the species conservation equations and Poisson's equation (i.e., so-called non-quasi-neutral approach) on the one hand and on a quasi-neutral approach on the other hand, but again these models were developed for argon.

For low-current nonthermal GA discharges (typically near 1 A or below), some simple 1D analytical or semi-analytical models have been developed [104–109], including the plasma string model [104] and the Elenbaas-Heller model, assuming an equilibrium plasma, with the radius of the plasma channel being constant [105–107] or with a correction based on an analytical relation between the electric field and the electron and gas temperatures for non-equilibrium plasma [108] or focusing on the discharge electrical parameters [109]. These simple models cannot describe the complex behavior of the GA, such as the unsteady behavior in time and space, arc restrike, non-equilibrium effects, effects of flow patterns, etc., and they did not include a detailed chemistry. Gutsol and Gangoli [110] presented a simple 2D model of a GA, in a plane parallel to the gas flow and perpendicular to the discharge current, which provided very useful information about the gas-discharge interaction. Within our group, we developed a 2D non-quasi-neutral fluid model for the arc gliding process in an argon GA [111], and we compared the glow and arc mode in this setup [112]. We also presented a 2D quasi-neutral model [113], which was also applied in 3D modeling for a classical (diverging electrode) GA [114] and a reverse vortex flow (RVF) GA (also called GA plasmatron; GAP) [115]. These models were developed for argon, but we also developed a 1D fluid model [44] and two different 2D models [86, 87] for a (classical or RVF) GA in CO₂, considering the detailed plasma chemistry of CO₂ conversion. An overview of both 0D chemical kinetic models and 2D/3D fluid models for plasma reactors of interest for CO₂ conversion was presented in [116].

3. Explanation of the modeling approaches

3.1 0D chemical kinetic modeling

Most models describing a detailed plasma chemistry apply the 0D chemical kinetic approach, which allows to handle a large number of species and chemical reactions, with limited computational effort. This approach is based on solving balance equations for the various species densities, based on production and loss rates, as defined by chemical reactions:

$$\frac{dn_i}{dt} = \sum_j \left\{ \left(a_{ij}^{(2)} - a_{ij}^{(1)} \right) k_j \prod_l n_l^{a_{ij}^{(1)}} \right\} \tag{1}$$

where $a_{ij}^{(1)}$ and $a_{ij}^{(2)}$ are the stoichiometric coefficients of species i, at the left- and right-hand sides of a reaction j, respectively, n_l is the species density at the left-hand side of the reaction, and k_j is the rate coefficient of reaction j. For the electron reactions, the energy-dependent rate coefficients are determined from the average electron energy, while the rate coefficients of the chemical reactions between neutral species or ions are adopted from the literature.

The species typically included in such models, for either pure CO_2 or pure CH_4, as well as the extra species included in CO_2/CH_4, CO_2/H_2O, CO_2/H_2, or CH_4/O_2 gas mixtures and in CO_2/N_2 and CH_4/N_2 mixtures are listed in **Table 1**. The same species can be included in the CO_2/CH_4, CO_2/H_2O, CO_2/H_2, and CH_4/O_2 models, because these combinations produce similar molecules. All the species listed in **Table 1** might chemically react with each other. Hence, a large number of chemical reactions (typically up to 1000) are incorporated in these models, including electron impact reactions, electron-ion recombination, and ion-ion, ion-neutral, and neutral-neutral reactions. Details of these chemistries for the specific gas mixtures can be found, for example, in [7, 28–30, 61–63, 66, 73–77, 83].

Specifically for CO_2 conversion in MW and GA plasmas, the vibrational levels of CO_2 are very important, because they allow energy-efficient CO_2 conversion [117], so the vibrational kinetics of CO_2 must be incorporated and especially the asymmetric stretch mode of CO_2, which is the most important channel for dissociation [117]. Likewise, in CO_2/N_2 mixtures in MW or GA discharges, the N_2 vibrational levels must be included, as they can populate the CO_2 vibrational levels [76]. Furthermore, also the vibrational levels of CO and O_2, and some electronically excited levels, are typically taken into account in such models (see **Table 1**). These vibrationally and electronically excited levels are indicated in **Table 1** with the symbols "V" and "E". Details about their notations can be found in [29, 76] or in **Table 1** for the N_2 electronically excited levels. Although vibrationally excited levels might also be important for CH_4, H_2O, and H_2 molecules, they are not yet included in the available models, to our knowledge, as these mixtures have only been described up to now for a DBD plasma, where the vibrational levels are of minor importance [117].

Although the above balance equations only account for time variations, thus neglecting spatial variations due to transport in the plasma, spatial variations can be included in such models, by imposing a certain input power or gas temperature as a function of time. For instance, this allows to account for microdischarge filaments in a DBD, through which the gas molecules pass when flowing through the reactor, by applying a number of pulses as a function of time (see, e.g., [28, 48, 62]). In a similar way, this method can account for the power deposition profile in a MW plasma (being at maximum at the position of the waveguide) by means of a temporal profile. Thus, the plasma reactors are considered as plug flow reactors, where

Molecules	Charged species	Radicals	Excited species
Species of interest in pure CO$_2$ models			
CO$_2$, CO	CO$_2^+$, CO$_4^+$, CO$^+$, C$_2$O$_2^+$, C$_2$O$_3^+$, C$_2$O$_4^+$, C$_2^+$, C$^+$, CO$_3^-$, CO$_4^-$	C$_2$O, C, C$_2$	CO$_2$(Va, Vb, Vc, Vd), CO$_2$(V1-V21), CO$_2$(E1, E2), CO(V1-V10), CO(E1-E4)
O$_2$, O$_3$	O$^+$, O$_2^+$, O$_4^+$, O$^-$, O$_2^-$, O$_3^-$, O$_4^-$	O	O$_2$(V1-V4), O$_2$(E1-E2)
	Electrons		
Species of interest in pure CH$_4$ models			
CH$_4$	CH$_5^+$, CH$_4^+$, CH$_3^+$, CH$_2^+$, CH$^+$, C$^+$	CH$_3$, CH$_2$, CH, C	CH$_4$*
C$_2$H$_6$, C$_2$H$_4$, C$_2$H$_2$, C$_2$	C$_2$H$_6^+$, C$_2$H$_5^+$, C$_2$H$_4^+$, C$_2$H$_3^+$, C$_2$H$_2^+$, C$_2$H$^+$, C$_2^+$	C$_2$H$_5$, C$_2$H$_3$, C$_2$H	C$_2$H$_6$*, C$_2$H$_4$*, C$_2$H$_2$*
C$_3$H$_8$, C$_3$H$_6$, C$_4$H$_2$		C$_3$H$_7$, C$_3$H$_5$	C$_3$H$_8$*
H$_2$	H$_3^+$, H$_2^+$, H$^+$, H$^-$	H	H$_2$*
Extra species typically included in CO$_2$/CH$_4$, CO$_2$/H$_2$O, CO$_2$/H$_2$ or CH$_4$/O$_2$ models			
H$_2$O, H$_2$O$_2$	H$_3$O$^+$, H$_2$O$^+$, OH$^+$, OH$^-$	OH, HO$_2$	H$_2$O*
CH$_2$O, CH$_3$OH, CH$_3$OOH		CHO, CH$_2$OH, CH$_3$O, CH$_3$O$_2$	
C$_2$H$_5$OH, C$_2$H$_5$OOH, CH$_3$CHO, CH$_2$CO		CHCO, CH$_3$CO, CH$_2$CHO, C$_2$H$_5$O, C$_2$H$_5$O$_2$	
Extra species typically included in CO$_2$/N$_2$ and/or CH$_4$/N$_2$ models			
N$_2$	N$^+$, N$_2^+$, N$_3^+$, N$_4^+$	N	N$_2$(V1-V14), N$_2$(C$^3\Pi_u$), N$_2$(A$^3\Sigma_u^+$), N$_2$(a$'^1\Sigma_u^-$), N$_2$(B$^3\Pi_g$), N(2D), N(2P)
N$_2$O, N$_2$O$_3$, N$_2$O$_4$, N$_2$O$_5$	NO$^+$, N$_2$O$^+$, NO$_2^+$, NO$^-$, N$_2$O$^-$, NO$_2^-$, NO$_3^-$, N$_2$O$_2^+$	NO, NO$_2$, NO$_3$	
HCN, ONCN, C$_2$N$_2$	HCN$^+$	H$_2$CN, CN, NCO, NCN	
NH$_3$	NH$_4^+$, NH$_3^+$, NH$_2^+$, NH$^+$	NH$_2$, NH	NH$_3$*
N$_2$H$_4$, N$_2$H$_2$		N$_2$H$_3$, N$_2$H	

Table 1.
Overview of the species typically included in plasma chemistry models for pure CO$_2$, pure CH$_4$, as well as extra species included in CO$_2$/CH$_4$, CO$_2$/H$_2$O, CO$_2$/H$_2$, and CH$_4$/O$_2$ gas mixtures and in CO$_2$/N$_2$ or CH$_4$/N$_2$ mixtures.

the plasma characteristics vary as a function of distance traveled by the gas, in the same way as they would vary in time in a batch reactor. The time in the balance equations thus corresponds to a residence time of the gas in the reactor, and the time variation can be translated into a spatial variation by means of the gas flow rate.

Besides balance equations for the species densities, 0D chemical kinetic models typically also apply balance equations for the electron temperature and/or the gas temperature, again based on source and loss terms, defined by the power deposition (or electric field) and the chemical reactions. Alternatively, instead of calculating the electron temperature with a balance equation, 0D models often solve a Boltzmann equation (e.g., Bolsig+ [118]), to calculate the electron energy distribution function (EEDF) and the rate coefficients of the electron impact reactions as a function of the electron energy. A more detailed description of the free electron kinetics in CO_2 plasma is provided in [32–37], where a state-to-state vibrational kinetic model was self-consistently coupled with the time-dependent electron Boltzmann equation.

0D models allow to predict the gas conversion, the product yields, and selectivities, based on the calculated plasma species densities at the beginning and the end of the simulations, corresponding to the inlet and outlet of the plasma reactor. Furthermore, based on the power introduced in the plasma and the gas flow rate, the specific energy input (SEI) can be computed, and from the latter, the energy efficiency (η) can be obtained with the following formulas:

$$SEI\left(\frac{kJ}{l}\right) = \frac{Plasma power(kW)}{Flowrate\left(\frac{l}{min}\right)} * 60\left(\frac{s}{min}\right) \qquad (2)$$

$$\eta(\%) = \frac{\Delta H_R\left(\frac{kJ}{mol}\right) * X_{CO_2}(\%)}{SEI\left(\frac{kJ}{l}\right) * 22.4\left(\frac{l}{mol}\right)} \qquad (3)$$

where ΔH_R is the reaction enthalpy of the reaction under study (e.g., 279.8 kJ/mol for CO_2 splitting) and X_{CO_2} is the CO_2 conversion. Note that this formula is only applicable to pure CO_2 splitting, but a similar formula can be applied to the other gas mixtures, using another reaction enthalpy and accounting not only for the CO_2 conversion but also for the conversion of the other gases in the mixture.

3.2 2D or 3D fluid modeling

Even though some spatial dependences of the plasma reactors can be taken into account in 0D chemical kinetic models, as explained above, they are not really suitable for describing detailed plasma reactor configuration or predict how modifications to the reactor geometry would give rise to better CO_2 conversion and energy efficiency. For this purpose, 2D or even 3D models are required, and fluid models are then the most logical choice, because they still allow a reasonable calculation time, in contrast to, for instance, PIC-MCC simulations.

These fluid models solve a number of conservation equations for the densities of the various plasma species and for the average electron energy. The energy of the other plasma species can be assumed in thermal equilibrium with the gas. The conservation equations for the species densities are again based on source and loss terms, defined by the chemical reactions, like in the 0D models. The source of the electron energy is due to heating by the electric field, and the energy loss is again dictated by collisions. In addition, transport is now included in the conservation equations, defined by diffusion and by migration in the electric field (for the charged species) and (in some cases) by convection due to the gas velocity. Furthermore, the conservation equations are coupled with Poisson's equation for a self-consistent calculation of the electric field distribution from the charged species densities, although more simplified quasi-neutral (QN) models have also been used [113], to further reduce the calculation time. Such a QN model neglects the near-electrode regions and treats only the quasi-neutral bulk plasma. It does not solve the Poisson equation, but calculates the ambipolar electric field from the ion densities and the electron and ion diffusion coefficients and mobilities.

Finally, in many cases, the gas temperature and gas flow behavior are calculated with a heat transfer equation and the Navier-Stokes equations, respectively, while in GA models, the cathode heat balance can also be accounted for, to properly describe the electron emission processes. The fluid (plasma) model and the models for gas flow and gas heating are typically combined into a multiphysics model: the calculated gas velocity is inserted in the transport equations of the plasma species, and the gas temperature determines the gas density profile and thus the chemical reaction rates.

4. Some typical calculation results

4.1 0D chemical kinetic modeling

0D chemical kinetic models typically provide information about the calculated gas conversion, energy efficiency, and product formation, as a function of specific operating conditions, as well as about the underlying chemistry explaining these results. The latter will be illustrated here, based on the modeling work performed within our group PLASMANT, for pure CO_2 splitting, as well as CO_2/CH_4, CH_4/O_2, CO_2/H_2, and CO_2/H_2O mixtures. For more details about the modeling results in these mixtures, and more specifically the calculated conversions, product yields and energy efficiencies, and comparison with experiments, we refer to the original research papers mentioned below, as well as two recent review papers [84, 116].

4.1.1 Pure CO_2 splitting

4.1.1.1 DBD conditions

The dominant reaction pathways for CO_2 splitting in a DBD plasma, as predicted from the model in [7], are plotted in **Figure 1**. As a DBD is characterized by relatively highly reduced electric field values (typically above 200 Td), and thus relatively high electron energies (several eV), electron impact reactions with CO_2 ground-state molecules dominate the chemistry. The most important reactions are electron impact dissociation into CO and O (which proceeds through electronically excited CO_2, that is, the so-called electron impact excitation-dissociation), electron impact ionization into CO_2^+ (which recombines with electrons or O_2^- ions into CO and O and/or O_2), and electron dissociative attachment into CO and O^- (cf. the thick black arrow lines in **Figure 1**). These three processes account for about 50%, 25%, and 25%, respectively, to the total CO_2 conversion [28]. Because these processes require more energy than strictly needed for breaking the C=O bond (i.e., 5.5 eV), the energy efficiency for CO_2 splitting in a DBD plasma is quite limited, that is, up to maximum 10% for a conversion up to 30% [1].

The CO molecules are relatively stable, but at very long residence time, they will recombine with O^- ions or O atoms, to form again CO_2 (cf. thin black arrow lines in **Figure 1**). This explains why the CO_2 conversion typically saturates at long residence times. Furthermore, the O atoms created upon CO_2 splitting also recombine quickly into O_2 or O_3, based on several processes (see also **Figure 1**).

4.1.1.2 MW and GA conditions

While our calculations predict that ca. 94% of the CO_2 splitting in a DBD plasma arises from the ground state, and only ~6% occurs from the vibrationally excited levels [28], the situation is completely different in a MW or GA plasma. These plasmas are characterized by much lower reduced electric field values (in the order of 50–100 Td), creating lower electron energies (order of 1 eV), which are most suitable for vibrational excitation of CO_2. Therefore, the CO_2 splitting in MW and GA discharge is mainly induced by electron impact vibrational excitation of the

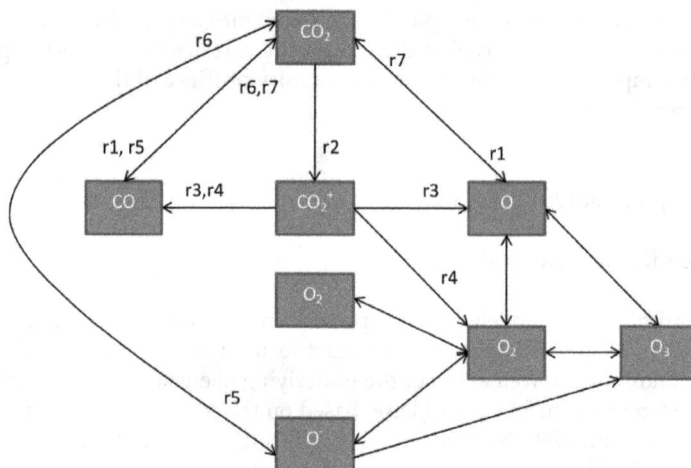

Figure 1.
Dominant reaction pathways of CO_2 splitting and the further reactions between O, O_2, and O_3 in a DBD plasma, as obtained from the model in [7], where the labels are also explained. Adopted from [119] with permission.

lowest vibrational levels, followed by vibrational-vibrational (VV) collisions, gradually populating the higher vibrational levels, leading to dissociation of CO_2. This stepwise vibrational excitation, or the so-called ladder climbing, is illustrated in **Figure 2**. As this process only requires 5.5 eV for dissociation, that is, exactly the C=O bond energy, this explains why MW and GA discharges exhibit a much better energy efficiency than a DBD, where the dominant dissociation mechanism is electron impact excitation-dissociation, as explained above, which requires 7–10 eV (see **Figure 2**).

Still, it must be realized that the vibrational excitation pathway is not always optimized in a MW or GA plasma. Indeed, as illustrated in detail in [42], the vibrational excitation is higher at lower pressures and higher power densities. The latter give rise to higher electron densities, which yield more vibrational excitation. Higher pressures, on the other hand, result in more vibrational-translational (VT) relaxation collisions, which represent the major loss mechanism of the vibrational energy. Finally, also the gas temperature plays a crucial role, as a higher gas temperature also results in more pronounced VT relaxation. Our models predict that in a MW plasma at atmospheric pressure, the dissociation is too much determined by thermal processes, thus limiting the CO_2 conversion and energy efficiency, in agreement with experimental observations. In addition, the recombination of CO and O atoms also becomes gradually more important at high gas temperature and pressures [42], further explaining why the experimental CO_2 conversion and energy efficiency drop upon increasing pressure. The main processes occurring in a MW plasma in the two extreme cases, that is, the ideal non-equilibrium conditions of low pressure and temperature and high power density and the near-thermal conditions of high pressure and temperature, are summarized in **Figure 3**. The model predicts a much higher CO_2 conversion and energy efficiency in a pressure range of 200–300 mbar and much lower values at atmospheric pressure, in the near-thermal conditions [42]. Hence, we should exploit as much as possible the non-equilibrium character of a MW plasma, in which the higher vibrational levels of CO_2 are overpopulated, to obtain the most energy-efficient CO_2 conversion.

The same conclusions can be drawn for a GA plasma, where our models predict that the CO_2 conversion could be further enhanced, by exploiting the role of the

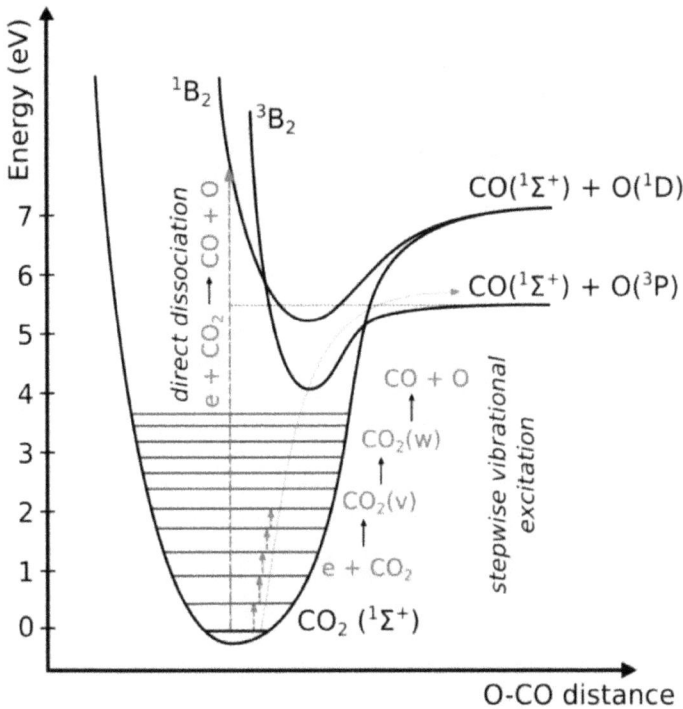

Figure 2.
Schematic illustration of some CO_2 electronic and vibrational levels, illustrating the energy-efficient dissociation process through electron impact vibrational excitation, followed by vibrational-vibrational collisions, which gradually populate the higher vibrational levels, that is, the so-called ladder climbing (5.5 eV), compared to direct dissociation through electronic excitation (above 7 eV). Adopted from [119] with permission.

higher vibrational levels of CO_2. Indeed, as the GA operates at atmospheric pressure, the vibrational distribution function (VDF) is too much thermal, that is, there is no significant overpopulation of the higher CO_2 vibrational levels. This was predicted both in a classical GA at a temperature around 1200 K [45] and in a RVF GA, operating at temperatures around 2500–3000 K [47]. The CO_2 dissociation even proceeds mainly from the ground state or the lowest vibrational levels. Indeed, based on these models, the major dissociation process was electron impact dissociation [45] or thermal dissociation [47] of the lower CO_2 vibrational levels, and the chemical reactions of the higher vibrational levels (with either O atoms or any arbitrary molecules in the plasma), which theoretically provide the most energy-efficient process for CO_2 conversion, were found to be of minor importance. Just like in the MW plasma, the model predicts that a significant overpopulation of the VDF, and thus a more energy-efficient CO_2 conversion, can be realized by decreasing the temperature or by increasing the power density [45].

4.1.2 CO_2/CH_4 mixture

When adding an H-source, such as CH_4, to the CO_2 plasma, a variety of molecules can be formed, with a mixture of H_2 and CO (or syngas) as the major compounds, but also smaller fractions of higher hydrocarbons and oxygenates can be formed. **Figure 4** illustrates the dominant pathways in a CO_2/CH_4 mixture, as predicted by the model in [66]. The thickness of the arrow lines is correlated to the rate of the reaction. CH_4 dissociation is initiated by electron impact, forming CH_3

a)

- High vibrational excitation + VV; low VT
- Dissociation by neutral and O collisions equally important
- Low recombination
- Energy efficient

b)

- Thermal VDF due to high VT
- High recombination
- Energy lost to heat

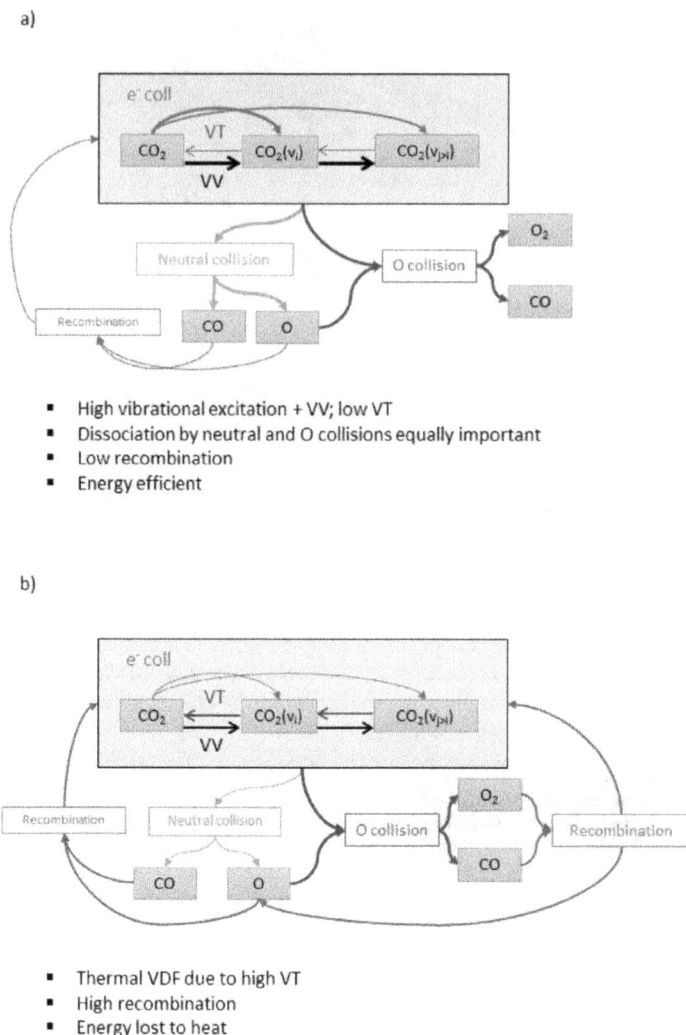

Figure 3.
Dominant reaction pathways of CO$_2$ splitting in a MW plasma, as obtained from the model in [42], for two extreme cases: (a) the ideal non-equilibrium conditions of low pressure and temperature and high power density and (b) the near-thermal condition of high pressure and temperature. Adopted from [116] with permission.

radicals, which recombine into higher hydrocarbons. Moreover, electron impact dissociation of CH$_4$ and of the higher hydrocarbons also yields H$_2$ formation. In addition, the CH$_3$ radicals also create methanol (CH$_3$OH) and CH$_3$O$_2$ radicals, albeit to a lower extent. Furthermore, the CH$_2$ radicals, also created from electron impact dissociation of CH$_4$, react with CO$_2$ to form formaldehyde (CH$_2$O) and CO. Finally, the O atoms, created from electron impact dissociation of CO$_2$ (see also **Figure 1**), also initiate the formation of higher oxygenates, like acetaldehyde (CH$_3$CHO). This species reacts further into CH$_3$CO radicals and subsequently into ketene (CH$_2$CO), although these pathways are not so important in absolute terms, as indicated by the thin dashed lines in **Figure 4**.

We have also compared the chemistry in the CO$_2$/CH$_4$ mixture, used for dry reforming of methane, with that of partial oxidation of methane, that is, a CH$_4$/O$_2$ mixture [66]. The reaction pathways of the latter are depicted in **Figure 5**. The CH$_4$/O$_2$ mixture clearly leads to a completely different chemistry than the CO$_2$/CH$_4$

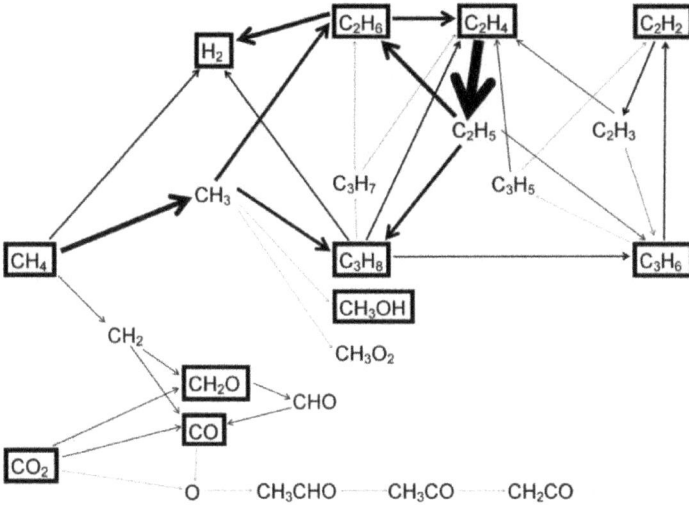

Figure 4.
Dominant reaction pathways for the conversion of CH_4 and CO_2 into higher hydrocarbons, H_2 and CO, and higher oxygenates, in a 70/30 CH_4/CO_2 DBD plasma, as obtained from the model in [66]. The thickness of the arrow lines corresponds to the importance of the reaction paths. Reproduced from [84] with permission.

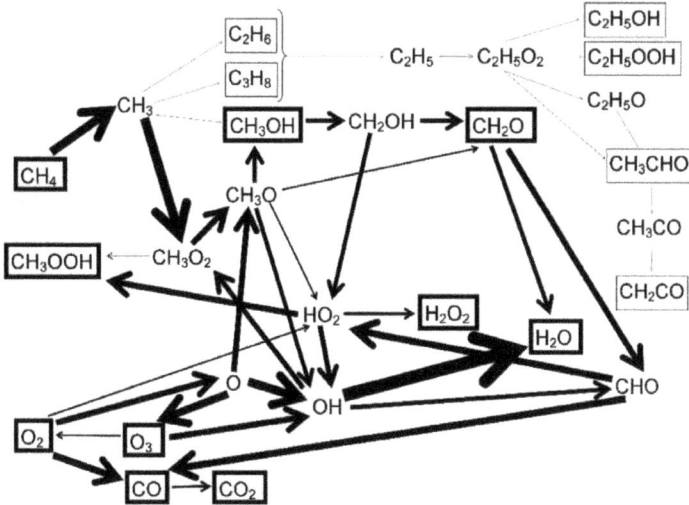

Figure 5.
Dominant reaction pathways for the conversion of CH_4 and O_2 into (mainly) higher oxygenates, as well as some full oxidation products, in a 70/30 CH_4/O_2 DBD plasma, as obtained from the model in [66]. The thickness of the arrow lines corresponds to the importance of the reaction paths. Reproduced from [84] with permission.

mixture, in spite of the fact that the same chemical species are included in the models (see **Table 1**). Electron impact dissociation of CH_4 again produces CH_3 radicals, which will recombine into methanol or higher hydrocarbons, but the recombination into CH_3O_2 radicals, which form either CH_3O radicals or methyl hydroperoxide (CH_3OOH), is now more important. The CH_3O radicals produce methanol, which seems a more important formation mechanism than the recombination of CH_3 with OH radicals (cf. the arrow line thickness in **Figure 5**), and methanol can also react further into CH_2OH radicals, producing formaldehyde. The latter is also easily converted into CHO radicals and further into CO (note the thickness of these arrow lines, indicating the importance of these reactions) and CO_2. Furthermore, formaldehyde is also partially converted into H_2O. Note that this pathway is illustrated for a 70/30 CH_4/O_2 mixture, which obviously leads to nearly full oxidation of CH_4, rather than partial oxidation, where the major end products should be the higher oxygenates. When less O_2 would be present in the mixture, our model predicts that methanol and methyl hydroperoxide are formed in nearly equal amounts as CO and H_2O [66]. **Figure 5** also illustrates that the O_2 molecules are mainly converted into CO, O atoms, and HO_2 radicals. Some O_3 is also formed out of O_2, but the reverse process, that is, the production of two O_2 molecules out of O_3 and O atoms, is more important, explaining why the arrow points from O_3 toward O_2. The O atoms are converted into CH_3O and OH radicals, producing methanol and water, respectively. The latter reaction (from OH to H_2O) appears to be very important (cf. thick arrow line in **Figure 5**), and thus, significant amounts of H_2O are formed, as predicted by the model [66].

In summary, comparing **Figures 4** and **5** clearly indicates that the chemical pathways in CH_4/O_2 and CH_4/CO_2 plasma are quite different, even at the same mixing ratios. Finally, in both mixtures a large number of different chemical compounds can be formed, but due to the reactivity of the plasma, there is no selective production of some targeted compounds. To reach the latter, the plasma will have to be combined with a catalyst.

4.1.3 CO_2/H_2 mixture

The dominant reaction pathways for the conversion of CO_2 and H_2 in a 50/50 CO_2/H_2 DBD plasma are illustrated in **Figure 6**, as predicted by the model in [75]. The conversion starts again with electron impact dissociation of CO_2, yielding CO and O atoms. Simultaneously, electron impact dissociation of H_2 results in the formation of H atoms, and this reaction seems more important (cf. the thickness of the arrow line). The O and H atoms recombine into the formation of OH radicals and further into H_2O. The model thus predicts that H_2O is produced at relatively high density [75]. The CO molecules will partially react back into CO_2, mainly through the formation of CHO radicals. This pathway appears to be more important than the direct three-body recombination between CO and O atoms into CO_2, which is the dominant pathway in a pure CO_2 plasma. The H atoms thus contribute significantly to the back reaction of CO into CO_2, and this explains why the

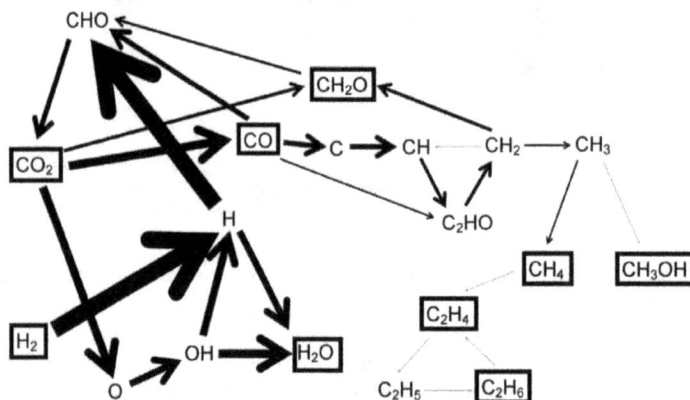

Figure 6.
Dominant reaction pathways for the conversion of CO_2 and H_2 into various products, in a 50/50 CO_2/H_2 DBD plasma, as obtained from the model in [75]. The thickness of the arrow lines corresponds to the rates of the net reactions. The stable molecules are indicated with black rectangles. Reproduced from [75] with permission.

calculated CO$_2$ conversion is quite limited in a CO$_2$/H$_2$ mixture [75]. Electron impact dissociation of CO yields the formation of C atoms, which react further into CH, CH$_2$, C$_2$HO, and CH$_3$ radicals in several successive radical recombination reactions. The CH$_2$ radicals react with CO$_2$ into CH$_2$O, while the CH$_3$ radicals easily form CH$_4$. The latter reaction is more favorable than CH$_3$OH formation out of CH$_3$. Finally, CH$_4$ partially reacts further into higher hydrocarbons (C$_x$H$_y$).

Figure 6 clearly illustrates that several subsequent radical reactions are required for the formation of (higher) hydrocarbons and oxygenates. This explains the very low calculated yields and selectivities of these end products [75]. In summary, the lack of direct formation of CH$_2$ and CH$_3$ in the CO$_2$/H$_2$ mixture, which is important in CO$_2$/CH$_4$ gas mixtures (see **Figure 4**), combined with the very low conversion of CO$_2$, which is again attributed to the absence of CH$_2$ as important collision partner for the loss of CO$_2$, makes the CO$_2$/H$_2$ mixture less interesting for the formation of higher hydrocarbons and oxygenates than a CO$_2$/CH$_4$ mixture at the conditions under study. Furthermore, as H$_2$ is a useful product by itself, while CH$_4$ is also a greenhouse gas (besides a fuel), the simultaneous conversion of CO$_2$ and CH$_4$, that is, two greenhouse gases, is considered to be of higher value, also because it represents a direct valorization of biogas.

4.1.4 CO$_2$/H$_2$O mixture

H$_2$O is the cheapest H-source to be added to a CO$_2$ plasma for the direct production of value-added chemicals, and the combined conversion of CO$_2$ and H$_2$O could mimic the natural photosynthesis process. However, adding H$_2$O (in concentrations up to 8%) to a CO$_2$ DBD plasma causes a significant reduction in the CO$_2$ conversion, while no oxygenated hydrocarbons were detected experimentally, and also the calculated concentrations were only in the ppb level [73].

These results can be explained by a kinetic analysis of the reaction chemistry. The latter reveals that the reaction between CO and OH, yielding H atoms and CO$_2$, is crucial, as it has a very high rate constant, and it controls the ratio between the conversions of CO$_2$ and H$_2$O. This can be explained in a very simple way by the following reactions:

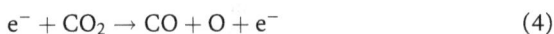

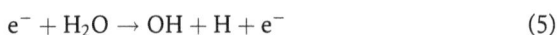

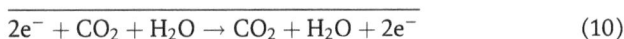

$$e^- + CO_2 \rightarrow CO + O + e^- \qquad (4)$$

$$e^- + H_2O \rightarrow OH + H + e^- \qquad (5)$$

$$CO + OH \rightarrow CO_2 + H \qquad (6)$$

$$H + O_2 + M \rightarrow HO_2 + M \qquad (7)$$

$$HO_2 + O \rightarrow OH + O_2 \qquad (8)$$

$$OH + H \rightarrow H_2O \qquad (9)$$

$$\overline{2e^- + CO_2 + H_2O \rightarrow CO_2 + H_2O + 2e^-} \qquad (10)$$

Reactions (4) and (5) yield the dissociation of CO$_2$ and H$_2$O, but the products, CO and OH, will rapidly recombine into CO$_2$ again (reaction (6)). Moreover, the two H atoms and one O atom formed will also quickly recombine, first into OH (through subsequent reactions (7) and (8)) and subsequently into H$_2$O through reaction (9). Thus, overall, there is no net dissociation of CO$_2$ and H$_2$O in this pathway (see overall reaction (10)).

Of course, there exist also other pathways for the conversion of these molecules, so there will still be some conversion of CO$_2$ and H$_2$O in the plasma, but electron impact dissociation is typically the major loss mechanism for CO$_2$ in a DBD (cf. also above), so the above mechanism explains the drop in CO_2 conversion upon addition of H_2O, as the OH radicals created upon H_2O dissociation give rise to the back reaction, creating CO_2 out of CO.

The above mechanism can also explain why no (significant) methanol (or other oxygenated hydrocarbons) is formed in the CO_2/H_2O mixture, because all the H atoms needed to form CH and CHO fragments for the formation of methanol are steered to OH and subsequently H_2O again. Hence, this chemical kinetic analysis indicates that H_2O might not be a suitable H-source for the formation of oxygenated hydrocarbons in a one-step process, because of the abundance of O atoms, O_2 molecules, and OH radicals, trapping the H atoms.

It should be noted that this fast reaction between H and O atoms was demonstrated to be useful for the O-trapping in the case of pure CO_2 conversion, thus providing a solution for the separation of the CO_2 splitting products [120], but in the present case, it is clearly the limiting factor for the formation of oxygenated hydrocarbons.

4.1.5 CO_2/N_2 mixture

Real industrial gas flows typically do not contain pure CO_2 but also other gases and impurities. In most cases, N_2 is the most important component. It is thus also important to study the effect of N_2 on the CO_2 conversion and energy efficiency, as well as which products are formed, that is, useful products or harmful NO_x compounds. Hence, we developed some models for a CO_2/N_2 mixture, both for a MW plasma [76] and a DBD [77]. Both models predict that N_2 has a beneficial effect on the CO_2 splitting, but the mechanism is completely different. In a DBD, the electronically excited metastable $N_2(A^3\Sigma_u^+)$ molecules give rise to the enhanced CO_2 splitting [77], while in a MW plasma, the N_2 vibrational levels help to populate the CO_2 vibrational levels, by VV relaxation, and this causes the enhanced CO_2 splitting [76]. It should be mentioned, however, that in spite of the higher absolute CO_2 conversion upon addition of N_2, the effective or overall CO_2 conversion will drop in both cases, because of the lower absolute fraction of CO_2 in the gas mixture. The effect is minor up to about 60% N_2 but more pronounced for higher N_2 fractions. As the effective CO_2 conversion determines the overall energy efficiency of the process, the latter also drops upon addition of more N_2, as some of the energy is used for ionization, excitation, and dissociation of the N_2 molecules.

Both the modeling and experiments also reveal that several NO_x compounds are produced in a CO_2/N_2 plasma, especially NO, NO_2, N_2O, and N_2O_5, as was discussed in detail in [77]. A detailed chemical kinetic analysis reveals how the NO_x compounds are formed and thus also how this formation can be reduced. As illustrated in **Figure 7**, N_2 is excited to a metastable state $N_2(A^3\Sigma_u^+)$, as well as dissociated into N atoms, by electron impact reactions. The $N_2(A^3\Sigma_u^+)$ molecules react with O atoms into NO or with O_2 into N_2O. The N atoms react with both O and O_3 into NO. NO can be converted into NO_2 upon reaction with O, but the opposite reaction, upon collision with either O or N atoms, occurs as well, making NO_2 the main source of NO production and vice versa (see **Figure 7**).

Furthermore, the N atoms are trapped in two reaction loops, that is, between NO, NO_2, and N_2O_3 and between NO_2, NO_3, and N_2O_5. The only way to escape from these loops is by the reaction of NO_2 to N_2O (which can react back to N_2 and N upon collision with $N_2(A^3\Sigma_u^+)$ and N_2^+) or by the reaction of NO with either N atoms or $N_2(a'^1\Sigma_u^-)$ molecules, forming again N atoms or N_2 molecules (see **Figure 7**).

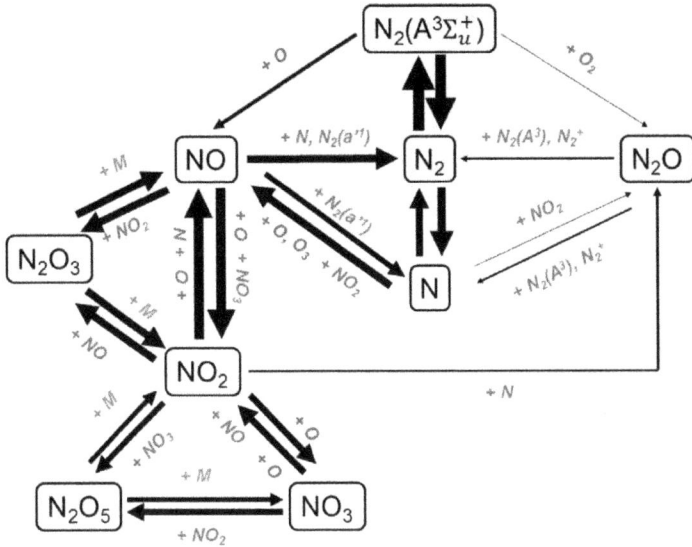

Figure 7.
Dominant reaction pathways leading to NO_x formation in a CO_2/N_2 DBD plasma, as obtained from the model in [77]. See details in the text. The thickness of the arrow lines corresponds to the time-integrated reaction rates, indicating the importance of the reactions. Adopted from [77] with permission.

Figure 7 shows that, in order to avoid the formation of NO_x compounds, we should prevent the reaction between the reactive N-species (i.e., $N_2(A^3\Sigma_u^+)$ and N) and the O species (O, O_2, or O_3). Reducing the concentrations of reactive N-species in the plasma is not straightforward, so we think that a more viable option to avoid NO_x formation is to remove the O atoms from the plasma, by means of O-scavengers, or separation membranes or a catalytic system.

4.2 2D or 3D fluid modeling

While 0D chemical kinetic models are most suitable to elucidate the underlying chemical reaction pathways, they cannot describe detailed effects of reactor design. For this purpose, fluid modeling is more appropriate. We will show here how 2D or 3D fluid models can help to obtain a better insight in the basic characteristics of plasma reactors, using two examples of our group PLASMANT, which are of great interest for the application of CO_2 conversion, that is, packed bed DBD reactors and reverse vortex flow GA reactors.

4.2.1 Packed bed DBD reactors

Packed bed DBD reactors are known to enhance the electric field and thus also the electron temperature, at the contact points between the packing pellets or beads, due to polarization of this dielectric packing. This is illustrated in **Figure 8**, showing the time-averaged electric field and electron temperature distributions in a 2D representation of a packed bed DBD reactor, for a peak-to-peak voltage of 4 kV and a frequency of 23.5 kHz. These results are obtained from a so-called "channel of voids" model, where the packing beads are not in direct contact, but allow the gas flowing through the packing. This is done to allow a 2D model representing a real 3D geometry (see details in [94]). In spite of the fact that there is no real contact between the beads, the local electric field enhancement in between the beads, due to

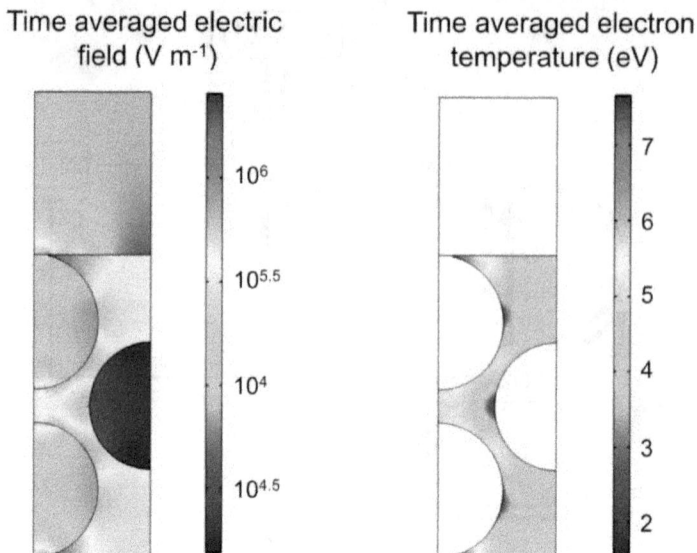

Figure 8.
Calculated time-averaged 2D profiles of the electric field and electron temperature in a packed bed DBD reactor, at a peak-to-peak voltage of 4 kV and a frequency of 23.5 kHz, as obtained from the model in [94]. Adopted from [94] with permission.

their polarization, is still visible, although it must be mentioned that the effect is more pronounced in a so-called "contact point" model (see [94]). This enhanced electric field gives rise to more electron heating and thus to a higher electron temperature in between the beads (see right panel of **Figure 8**). At this relatively low applied voltage of 4 kV, the plasma is initiated at the contact points and remains in this region, reflecting the properties of a Townsend discharge, while at higher applied voltage, for example, 7.5 kV (peak-to-peak), the discharge will spread out more into the bulk of the reactor, from one void space to the other, ultimately covering the whole gas gap [94]. Such behavior was also reported from experiments. Indeed, by means of an intensified charge-coupled device (ICCD) camera [121, 122], Kim and coworkers also observed that at low applied potential, the discharge stays local at the contact points, while at higher potential, it spreads across the surface of the packing material, and similar observations were also made by Tu et al. [123].

Although the above model was developed for helium, we expect a similar behavior in a CO_2 plasma. The higher electron temperature will give rise to more electron impact ionization, excitation, and dissociation of the CO_2 molecules, for the same applied power, and this can explain why a packed bed DBD gives a higher CO_2 conversion and energy efficiency than an empty reactor.

We also developed a model for a packed bed DBD reactor in dry air, to study the propagation of a plasma streamer [97], as illustrated in **Figures 9** and **10**. Our calculations reveal that the plasma formation in a packed bed DBD reactor in dry air may exhibit three types of discharge behavior, that is, positive restrikes; filamentary microdischarges, also localized between the packing beads; and surface ionization waves, in agreement with the model by Kruszelnicki et al. [93]. Positive restrikes between the dielectrics result in the formation of filamentary microdischarges. Surface charging creates electric field components parallel to the dielectric surface and leads to the formation of surface ionization waves. At a low

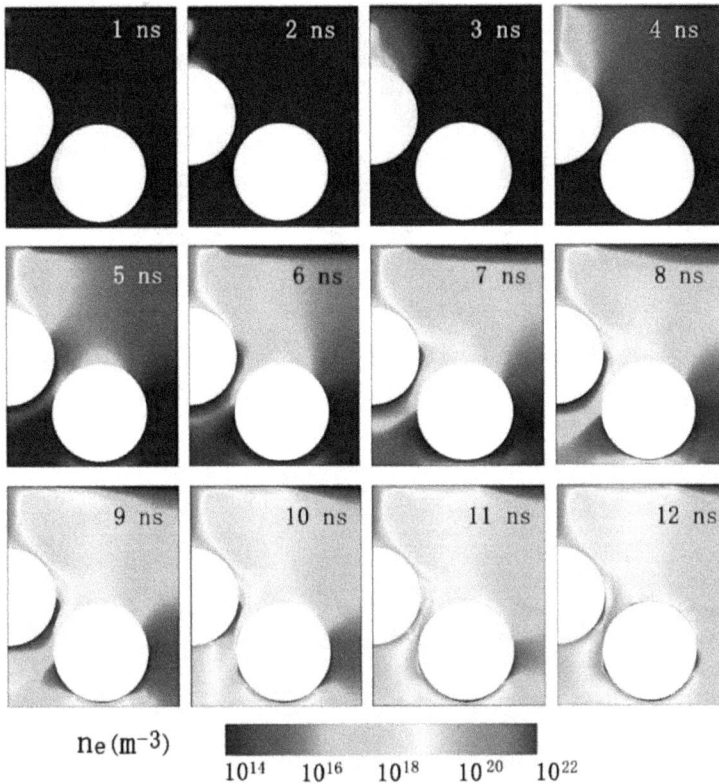

Figure 9.
Calculated electron number density distribution as a function of time, for a packed bed DBD reactor in dry air, with packing beads of $\varepsilon_r = 5$. Adopted from [97] with permission.

dielectric constant of the packing ($\varepsilon_r = 5$), plasma ignition between the beads occurs directly in the mode of surface discharges (or surface ionization waves), which can connect with the surface of the adjacent bead; see **Figure 9**. On the other hand, at high dielectric constants ($\varepsilon_r = 1000$), no surface streamer jumping toward the adjacent bead surface is observed, and spatially limited filamentary microdischarges, so-called local discharges, are generated between the beads; see **Figure 10**. For intermediate dielectric constants, a mixed mode of surface discharges and local discharges exists [97]. Good qualitative agreement with experiments was obtained, as detailed in [97].

The positive restrikes, local discharges, and surface discharges all give rise to the production of reactive species, because they exhibit an enhanced electric field and thus they create a burst of energetic electrons, which produce reactive species by electron impact dissociation. Packed bed reactors are often used for plasma catalysis, where packing beads with different dielectric constants can act as supports for the catalytic materials. Therefore, this study is important to gain a better insight on how different packing materials can influence the performance of packed bed DBD reactors for plasma catalysis. As our results indicate that a higher dielectric constant constrains the discharge to the contact points of the beads, this may limit the catalyst activation due to the limited catalyst surface area in contact with the discharge, and thus it may have implications for the efficiency of plasma catalytic CO₂ conversion. Indeed, the best results are not always reached for the highest dielectric constant [9, 10].

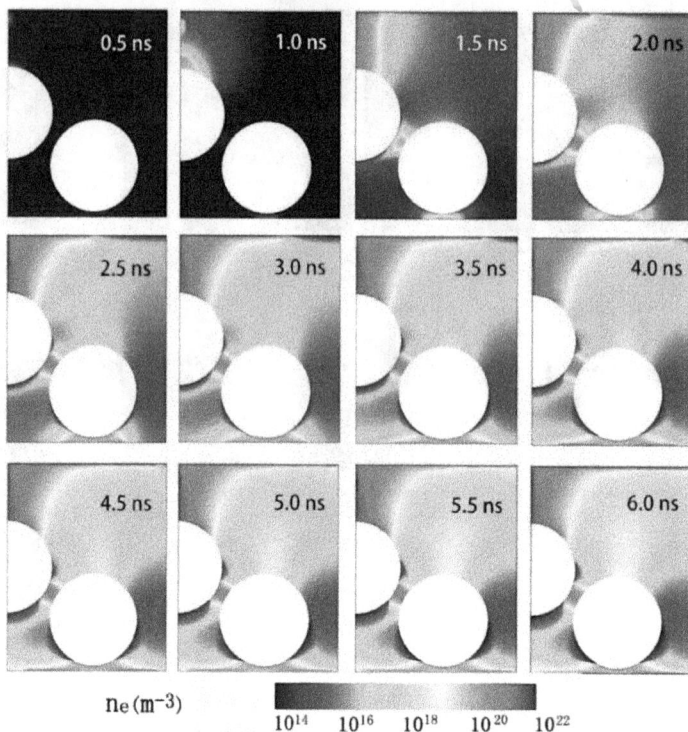

$n_e\,(m^{-3})$

10^{14} 10^{16} 10^{18} 10^{20} 10^{22}

Figure 10.
Calculated electron number density distribution as a function of time, for a packed bed DBD reactor in dry air, with packing beads of ε_r = 1000. Adopted from [97] with permission.

4.2.2 Gliding arc plasmatron (GAP)

Figure 11 illustrates a typical 3D gas flow pattern (a), as well as the calculated electron density profile (b), in a reverse vortex flow (RVF) GA plasma reactor, also called gliding arc plasmatron (GAP), operating in argon. The stream line plot clearly

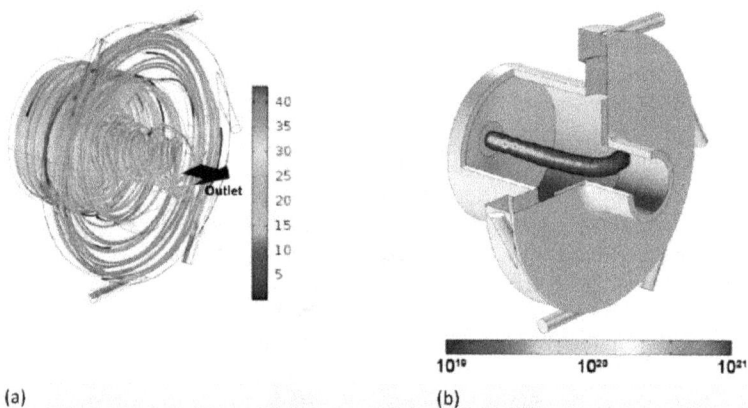

(a)   (b)

Figure 11.
Calculated steady-state gas flow stream lines (a) and electron density at a time of 5.3 ms, when the arc is stabilized in the center (b), for a reverse vortex flow GA plasma reactor at an arc current of 240 mA, as obtained from the model in [87].

depicts the formation of a reverse vortex flow. The gas is forced into a tangential motion due to the tangential inlets and travels in this way, close to the sidewalls, toward the closed cathode side at the end (= back of **Figure 11(a)**) with a velocity around 30–40 m/s. After it has reached the closed cathode end, it moves in the opposite direction, in a smaller inner (reverse) vortex toward the outlet, with much lower velocity, and it exits the reactor with a velocity around 20 m/s (see also the color scale in **Figure 11(a)**).

The arc plasma reacts to this gas flow pattern, in the sense that when the gas stream is forced to the center, the arc channel will also move to the center (due to convection), and it will stay in this position as long as the gas keeps it stabilized. Hence, the arc plasma is effectively stabilized in the center, as illustrated in **Figure 11(b)**. Furthermore, as the mass transfer is directed toward the center, the walls are thermally insulated from the hot plasma arc column. The fact that no heat is lost to the reactor walls or other parts of the reactor means that more power can be consumed by the discharge, that is, the plasma generation is more effective. Furthermore, keeping the walls insulated (cold) is also beneficial for the reactor materials itself. The calculated plasma density, for an arc current of 240 mA, is around 10^{20} m^{-3}, which is a typical value for GA plasmas at atmospheric pressure.

It should be noted that the results of **Figure 11** are for an argon plasma, but we also developed a similar model for a CO_2 plasma, but this was only possible in 2D, because of computation time [87]. However, in the same paper, we also "downgraded" the 3D argon model into 2D, and a comparison between both indicated that the difference between the 3D and 2D argon models was limited. Therefore, the 2D CO_2 model provides data with reasonable accuracy. We calculated a typical plasma density of 4×10^{19} m^{-3} in the arc center, which is about one order of magnitude lower than in argon (4×10^{20} m^{-3}), due to other chemical processes, not all leading to ionization.

We can conclude from **Figure 11** that the gas, when moving in the inner vortex flow will largely pass through the arc column. This result is very interesting for the application of CO_2 conversion, as it shows that the design of this GAP allows more gas to pass through the arc zone than in a classical (diverging electrodes) GA. Nevertheless, our combined simulations and experiments reveal that the fraction of gas that passes through the arc is still somewhat limited, thus limiting the overall CO_2 conversion [19, 20]. By means of this type of 3D fluid dynamics modeling, we aim to predict a more optimized design, to further improve the application of CO_2 conversion.

5. Conclusion

Plasma-based CO_2 conversion is gaining increasing interest, but to improve this application, we need to obtain a better insight in the underlying mechanisms. The latter can be obtained by both plasma chemistry modeling and plasma reactor modeling. This chapter shows some examples of both modeling approaches from our own group, to illustrate what type of information can be obtained from such models and how this modeling can contribute to a better insight, in order to improve this application.

0D chemical reaction kinetic modeling is very suitable for describing the underlying plasma chemical reaction pathways of the conversion process. We have illustrated this for pure CO_2 splitting, showing the difference between a DBD and MW/GA plasma. Indeed, in a DBD, the CO_2 conversion is mainly due to electron impact electronic excitation followed by dissociation with the CO_2 ground-state molecules, which requires about 7–10 eV per molecule. This "waste of energy" explains the lower energy efficiency of CO_2 splitting in a DBD. On the other hand, in a MW and GA plasma, vibrational excitation of CO_2 is dominant, and VV relaxation gradually populates the higher vibrational levels (so-called ladder climbing). This is the most energy-efficient way of CO_2 dissociation, as it requires only 5.5 eV per molecule, that is, exactly the C=O bond energy.

We also presented the important reaction pathways in CO_2/CH_4, CH_4/O_2, CO_2/H_2 and CO_2/H_2O mixtures, as well as for the effect of N_2 addition to a CO_2 plasma. In a DBD plasma, the conversion is always initiated by electron impact dissociation, creating radicals that react further into value-added compounds. The main products formed are syngas (CO/H_2), but higher hydrocarbons and oxygenates are also formed in limited amounts. However, the selective production of these targeted compounds is not yet possible, due to the high reactivity of the plasma. Therefore, a catalyst must be inserted in the plasma. Our models reveal that CO_2/CH_4 and CH_4/O_2 mixtures exhibit totally different chemical reactions, resulting in different products. A CO_2/H_2 mixture does not produce many higher hydrocarbons and oxygenates, and the CO_2 conversion is very limited, due to the lack of CH_2 (and CH_3) radical formation. Indeed, the CH_2 radicals are the main collision partners of CO_2 in the CO_2/CH_4 mixture. Furthermore, adding H_2O to a CO_2 DBD plasma yields a drop in CO_2 conversion, and also the H_2O conversion is limited, and virtually no oxygenated hydrocarbons are formed, which could also be explained from the chemical reaction paths. The insights obtained by the model might be useful to provide possible solutions. The last example of 0D chemical kinetic modeling was given for a CO_2/N_2 plasma, where it was shown that also NO_x compounds are produced, which might give several environmental problems. Again, the model can explain their formation, which is useful to provide possible solutions on how to avoid this NO_x formation.

Although 0D models can give useful information on the plasma chemistry, they cannot really account for details in the plasma reactor configuration and thus predict how modifications to the reactor design might lead to improved CO_2 conversion. For this purpose, 2D or 3D fluid models of specific reactor designs are needed. Developing such fluid models for a detailed plasma chemistry, however, leads to excessive calculation times. Therefore, these models are up to now mainly developed for simpler chemistry, in argon or helium. We have shown here examples for a packed bed DBD reactor and a GAP. These models allow to elucidate why certain reactor designs give beneficial results and to pinpoint the limitations and finally how improvements in the reactor designs might yield a better CO_2 conversion and energy efficiency.

In the future work, we intend to implement the more complex CO_2 chemistry (either pure or mixed with other gases) in such fluid models, to obtain a more comprehensive picture of CO_2 conversion in a real plasma reactor geometry. As this is quite challenging in terms of computation time, reduced chemistry sets must be developed for CO_2 and its gas mixtures. When modeling CO_2 conversion in a MW or GA plasma, the vibrational kinetics must be accounted for. To avoid the need of describing all individual levels, we have developed a level-lumping strategy [39], which enables to group the vibrational levels of the asymmetric stretch mode of CO_2 into a number of groups. This reduces the calculation time, so that it can be implemented in 2D models [86]. We believe that a combination of 0D chemical kinetic models (to obtain detailed insight in the entire plasma chemistry and to develop reduced chemistry sets, identifying the main species and chemical reactions) and 2D/3D fluid models (for a detailed understanding of the reactor design) is the most promising approach to make further progress in this field.

Acknowledgements

We would like to thank R. Aerts, A. Berthelot, C. De Bie, T. Kozák, and K. Van Laer for sharing their simulation results.

Author details

Annemie Bogaerts[1*], Ramses Snoeckx[1,2], Georgi Trenchev[1] and Weizong Wang[1,3]

1 Department of Chemistry, Research Group PLASMANT, University of Antwerp, Belgium

2 Clean Combustion Research Center (CCRC), Physical Science and Engineering Division (PSE), King Abdullah University of Science and Technology (KAUST), Saudi Arabia

3 School of Astronautics, Beihang University, Beijing, P.R. China

*Address all correspondence to: annemie.bogaerts@uantwerpen.be

IntechOpen

References

[1] Snoeckx R, Bogaerts A. Plasma technology - a novel solution for CO_2 conversion? Chemical Society Reviews. 2017;**46**:5805-5863

[2] Paulussen S, Verheyde B, Tu X, De Bie C, Martens T, Petrovic D, et al. Conversion of carbon dioxide to value-added chemicals in atmospheric pressure dielectric barrier discharges. Plasma Sources Science and Technology. 2010;**19**:034015

[3] Tu X, Gallon HJ, Twigg MV, Gorry PA, Whitehead JC. Dry reforming of methane over a Ni/Al_2O_3 catalyst in a coaxial dielectric barrier discharge reactor. Journal of Physics D: Applied Physics. 2011;**44**:274007

[4] Gómez-Ramírez A, Rico VJ, Cotrino J, González-Elipe AR, Lambert RM. Low temperature production of formaldehyde from carbon dioxide and ethane by plasma-assisted catalysis in a ferroelectrically moderated dielectric barrier discharge reactor. ACS Catalysis. 2014;**4**:402-408

[5] Scapinello M, Martini LM, Tosi P. CO_2 hydrogenation by CH_4 in a dielectric barrier discharge: Catalytic effect of Ni and Cu. Plasma Processes and Polymers. 2014;**11**:624-628

[6] Ozkan A, Dufour T, Arnoult G, De Keyzer P, Bogaerts A, Reniers F. CO_2-CH_4 conversion and syngas formation at atmospheric pressure using a multi-electrode dielectric barrier discharge. Journal of CO_2 Utilization. 2015;**9**:74-81

[7] Aerts R, Somers W, Bogaerts A. Carbon dioxide splitting in a dielectric barrier discharge plasma: A combined experimental and computational study. ChemSusChem. 2015;**8**:702-716

[8] Van Laer K, Bogaerts A. Improving the conversion and energy efficiency of carbon dioxide splitting in a zirconia-packed dielectric barrier discharge reactor. Energy Technology. 2015;**3**:1038-1044

[9] Michielsen I, Uytdenhouwen Y, Pype J, Michielsen B, Mertens J, Reniers F, et al. CO_2 dissociation in a packed bed DBD reactor: First steps towards a better understanding of plasma catalysis. Chemical Engineering Journal. 2017;**326**:477-488

[10] Uytdenhouwen Y, Van Alphen S, Michielsen I, Meynen V, Cool P, Bogaerts A. A packed-bed DBD micro plasma reactor for CO_2 dissociation: Does size matter? Chemical Engineering Journal. 2018;**348**:557-568

[11] Silva T, Britun N, Godfroid T, Snyders R. Optical characterization of a microwave pulsed discharge used for dissociation of CO_2. Plasma Sources Science and Technology. 2014;**23**:025009

[12] Bongers W, Bouwmeester H, Wolf B, Peeters F, Welzel S, van den Bekerom D, et al. Plasma-driven dissociation of CO_2 for fuel synthesis. Plasma Processes and Polymers. 2017;**16**:1600126

[13] van Rooij GJ, van den Bekerom DCM, den Harder N, Minea T, Berden G, Bongers WA, et al. Taming microwave plasma to beat thermodynamics in CO_2 dissociation. Faraday Discussions. 2015;**183**:233-248

[14] Scapinello M, Martini LM, Dilecce G, Tosi P. Conversion of CH_4/CO_2 by a nanosecond repetitively pulsed discharge. Journal of Physics D: Applied Physics. 2016;**49**:075602

[15] Zhu B, Li XS, Liu JL, Zhu X, Zhu AM. Kinetics study on carbon dioxide reforming of methane in kilohertz

spark-discharge plasma. Chemical Engineering Journal. 2015;**264**:445-452

[16] Nunnally T, Gutsol K, Rabinovich A, Fridman A, Gutsol A, Kemoun A. Dissociation of CO_2 in a low current gliding arc plasmatron. Journal of Physics D: Applied Physics. 2011;**44**: 274009

[17] Tu X, Whitehead JC. Plasma dry reforming of methane in an atmospheric pressure AC gliding arc discharge: Co-generation of syngas and carbon nanomaterials. International Journal of Hydrogen Energy. 2014;**39**:9658-9669

[18] Liu JL, Park HW, Chung WJ, Ahn WS, Park DW. Simulated biogas oxidative reforming in AC-pulsed gliding arc discharge. Chemical Engineering Journal. 2016;**285**:243-251

[19] Ramakers M, Trenchev G, Heijkers S, Wang W, Bogaerts A. Gliding arc plasmatron: Providing an alternative method for carbon dioxide conversion. ChemSusChem. 2017;**10**:2642-2652

[20] Cleiren E, Heijkers S, Ramakers M, Bogaerts A. Dry reforming of methane in a gliding arc plasmatron: Towards a better understanding of the plasma chemistry. ChemSusChem. 2017;**10**: 4025-4036

[21] Neyts EC, Ostrikov K, Sunkara MK, Bogaerts A. Plasma catalysis: Synergistic effects at the nanoscale. Chemical Reviews. 2015;**115**:13408-13446

[22] Hokazono H, Fujimoto H. Theoretical analysis of the CO_2 molecule decomposition and contaminants yield in transversely excited atmospheric CO_2 laser discharge. Journal of Applied Physics. 1987;**62**:1585-1594

[23] Cenian A, Chernukho A, Borodin V, Sliwinski G. Modeling of plasma-chemical reactions in gas mixture of CO_2 lasers I. gas decomposition in pure CO_2 glow discharge. Contributions to Plasma Physics. 1994;**34**:25-37

[24] Cenian A, Chernukho A, Borodin V. Modeling of plasma-chemical reactions in gas mixture of CO_2 lasers. II. Theoretical model and its verification. Contributions to Plasma Physics. 1995; **35**:273-296

[25] Gordiets BF, Osipov AI, Stupochenko EV, Shelepin LA. Vibrational relaxation in gases and molecular lasers. Soviet Physics Uspekhi. 1973;**15**:759-785

[26] Kustova E, Nagnibeda E. On a correct description of a multi-temperature dissociating CO_2 flow. Chemical Physics. 2006;**321**:293-310

[27] Rusanov VD, Fridman AA, Sholin GV. The physics of a chemically active plasma with nonequilibrium vibrational excitation of molecules. Soviet Physics Uspekhi. 1981;**24**:447-474

[28] Aerts R, Martens T, Bogaerts A. Influence of vibrational states on CO_2 splitting by dielectric barrier discharges. Journal of Physical Chemistry C. 2012; **116**:23257-23273

[29] Kozák T, Bogaerts A. Splitting of CO_2 by vibrational excitation in non-equilibrium plasmas: A reaction kinetics model. Plasma Sources Science and Technology. 2014;**23**:045004

[30] Kozák T, Bogaerts A. Evaluation of the energy efficiency of CO_2 conversion in microwave discharges using a reaction kinetics model. Plasma Sources Science and Technology. 2015;**24**: 015024

[31] Ponduri S, Becker MM, Welzel S, van de Sanden MCM, Loffhagen D, Engeln R. Fluid modelling of CO_2 dissociation in a dielectric barrier discharge. Journal of Applied Physics. 2016;**119**:093301

[32] Pietanza LD, Colonna G, D'Ammando G, Laricchiuta A, Capitelli M. Vibrational excitation and dissociation mechanisms of CO_2 under non-equilibrium discharge and post-discharge conditions. Plasma Sources Science and Technology. 2015;**24**: 042002

[33] Pietanza LD, Colonna G, D'Ammando G, Laricchiuta A, Capitelli M. Non equilibrium vibrational assisted dissociation and ionization mechanisms in cold CO_2 plasmas. Chemical Physics. 2016;**468**:44-52

[34] Pietanza LD, Colonna G, D'Ammando G, Laricchiuta A, Capitelli M. Electron energy distribution functions and fractional power transfer in "cold" and excited CO_2 discharge and post discharge conditions. Physics of Plasmas. 2016;**23**:013515

[35] Pietanza LD, Colonna G, Laporta V, Celiberto R, D'Ammando G, Laricchiuta A, et al. Influence of electron molecule resonant vibrational collisions over the symmetric mode and direct excitation-dissociation cross sections of CO_2 on the electron energy distribution function and dissociation mechanisms in cold pure CO_2 plasmas. The Journal of Physical Chemistry. A. 2016;**120**: 2614-2628

[36] Pietanza LD, Colonna G, D'Ammando G, Capitelli M. Time-dependent coupling of electron energy distribution function, vibrational kinetics of the asymmetric mode of CO_2 and dissociation, ionization and electronic excitation kinetics under discharge and post-discharge conditions. Plasma Physics and Controlled Fusion. 2016;**59**:014035

[37] Koelman P, Heijkers S, Tadayon Mousavi S, Graef W, Mihailova D, Kozák T, et al. A comprehensive chemical model for the splitting of CO_2 in non-equilibrium plasmas. Plasma

Processes and Polymers. 2017;**14**: 1600155

[38] Chen JL, Wang HX, Sun SR. Analysis of dissociation mechanism of CO_2 in a micro-hollow cathode discharge. Chinese Physics Letters. 2016;**33**:108201

[39] Berthelot A, Bogaerts A. Modeling of plasma-based CO_2 conversion: Lumping of the vibrational levels. Plasma Sources Science and Technology. 2016;**25**:045022

[40] de la Fuente JF, Moreno SH, Stankiewicz AI, Stefanidis GD. A new methodology for the reduction of vibrational kinetics in non-equilibrium microwave plasma: Application to CO_2 dissociation. Reaction Chemistry & Engineering. 2016;**1**:540-554

[41] Moss M, Yanallah K, Allen R, Pontiga F. An investigation of CO_2 splitting using nanosecond pulsed corona discharge: Effect of argon addition on CO_2 conversion and energy efficiency. Plasma Sources Science and Technology. 2017;**26**:035009

[42] Berthelot A, Bogaerts A. Modeling of CO_2 splitting in a microwave plasma: How to improve the conversion and energy efficiency. Journal of Physical Chemistry C. 2017;**121**:8236-8251

[43] Berthelot A, Bogaerts A. Pinpointing energy losses in CO_2 plasmas - effect on CO_2 conversion. Journal of CO_2 Utilization. 2018;**24**: 479-499

[44] Wang W, Berthelot A, Kolev S, Tu X, Bogaerts A. CO_2 conversion in a gliding arc plasma: 1D cylindrical discharge model. Plasma Sources Science and Technology. 2016;**25**: 065012

[45] Sun SR, Wang HX, Mei DH, Tu X, Bogaerts A. CO_2 conversion in a gliding arc plasma: Performance improvement

based on chemical reaction modeling. Journal of CO_2 Utilization. 2017;**17**: 220-234

[46] Berthelot A, Bogaerts A. Modeling of CO_2 plasma: Effect of uncertainties in the plasma chemistry. Plasma Sources Science and Technology. 2017;**26**: 115002

[47] Heijkers S, Bogaerts A. CO_2 conversion in a gliding arc plasmatron: Elucidating the chemistry through kinetic modeling. Journal of Physical Chemistry C. 2017;**121**:22644-22655

[48] Bogaerts A, Wang W, Berthelot A, Guerra V. Modeling plasma-based CO_2 conversion: Crucial role of the dissociation cross section. Plasma Sources Science and Technology. 2016; **25**:055016

[49] Indarto A, Choi JW, Lee H, Song HK. Kinetic modeling of plasma methane conversion using gliding arc. Journal of Natural Gas Chemistry. 2005; **14**:13-21

[50] Indarto A, Coowanitwong N, Choi JW, Lee H, Song HK. Kinetic modeling of plasma methane conversion in a dielectric barrier discharge. Fuel Processing Technology. 2008;**89**: 214-219

[51] Yang Y. Direct non-oxidative methane conversion by non-thermal plasma: Modeling study. Plasma Chemistry and Plasma Processing. 2003; **23**:327

[52] De Bie C, Verheyde B, Martens T, van Dijk J, Paulussen S, Bogaerts A. Fluid modelling of the conversion of methane into higher hydrocarbons in an atmospheric pressure dielectric barrier discharge. Plasma Processes and Polymers. 2011;**8**:1033-1058

[53] Luche J, Aubry O, Khacef A, Cormier JM. Syngas production from methane oxidation using a non-thermal plasma: Experiments and kinetic modeling. Chemical Engineering Journal. 2009;**149**:35-41

[54] Zhou LM, Xue B, Kogelschatz U, Eliasson B. Nonequilibrium plasma reforming of greenhouse gases to synthesis gas. Energy & Fuels. 1998;**12**: 1191-1199

[55] Machrafi H, Cavadias S, Amouroux J. CO_2 valorization by means of dielectric barrier discharge. Journal of Physics: Conference Series. 2011;**275**: 012016

[56] Goujard V, Tatibouet JM, Batiot-Dupeyrat C. Carbon dioxide reforming of methane using a dielectric barrier discharge reactor: Effect of helium dilution and kinetic model. Plasma Chemistry and Plasma Processing. 2011; **31**:315-325

[57] Wang JG, Liu CJ, Eliasson B. Density functional theory study of synthesis of oxygenates and higher hydrocarbons from methane and carbon dioxide using cold plasmas. Energy & Fuels. 2004;**18**: 148-153

[58] Istadi I, Amin NAS. Modelling and optimization of catalytic–dielectric barrier discharge plasma reactor for methane and carbon dioxide conversion using hybrid artificial neural network— Genetic algorithm technique. Chemical Engineering Science. 2007;**62**:6568-6581

[59] Kraus M, Egli W, Haffner K, Eliasson B, Kogelschatz U, Wokaun A. Investigation of mechanistic aspects of the catalytic CO_2 reforming of methane in a dielectric-barrier discharge using optical emission spectroscopy and kinetic modeling. Physical Chemistry Chemical Physics. 2002;**4**:668-675

[60] Liu CJ, Li Y, Zhang YP, Wang Y, Zou J, Eliasson B, et al. Production of acetic acid directly from methane and carbon dioxide using dielectric-barrier

discharges. Chemistry Letters. 2001;**30**: 1304-1305

[61] De Bie C, Martens T, van Dijk J, Paulussen S, Verheyde B, Bogaerts A. Dielectric barrier discharges used for the conversion of greenhouse gases: Modeling the plasma chemistry by fluid simulations. Plasma Sources Science and Technology. 2011;**20**:024008

[62] Snoeckx R, Aerts R, Tu X, Bogaerts A. Plasma-based dry reforming: A computational study ranging from the nanoseconds to seconds time scale. Journal of Physical Chemistry C. 2013; **117**:4957-4970

[63] Snoeckx R, Zeng YX, Tu X, Bogaerts A. Plasma-based dry reforming: Improving the conversion and energy efficiency in a dielectric barrier discharge. RSC Advances. 2015;**5**: 29799-29808

[64] Janeco A, Pinhão NR, Guerra V. Electron kinetics in He/CH$_4$/CO$_2$ mixtures used for methane conversion. Journal of Physical Chemistry C. 2015; **119**:109-120

[65] Wang W, Berthelot A, Zhang QZ, Bogaerts A. Modelling of plasma-based dry reforming: How do uncertainties in the input data affect the calculation results? Journal of Physics D: Applied Physics. 2018;**51**:204003

[66] De Bie C, van Dijk J, Bogaerts A. The dominant pathways for the conversion of methane into oxygenates and syngas in an atmospheric pressure dielectric barrier discharge. Journal of Physical Chemistry C. 2015;**119**: 22331-22350

[67] Zhou LM, Xue B, Kogelschatz U, Eliasson B. Partial oxidation of methane to methanol with oxygen or air in a nonequilibrium discharge plasma. Plasma Chemistry and Plasma Processing. 1998;**18**:375-393

[68] Nair SA, Nozaki T, Okazaki K. Methane oxidative conversion pathways in a dielectric barrier discharge reactor —Investigation of gas phase mechanism. Chemical Engineering Journal. 2007;**132**:85-95

[69] Goujard V, Nozaki T, Yuzawa S, Ağiral A, Okazaki K. Plasma-assisted partial oxidation of methane at low temperatures: Numerical analysis of gas-phase chemical mechanism. Journal of Physics D: Applied Physics. 2011;**44**: 274011

[70] Agiral A, Nozaki T, Nakase M, Yuzawa S, Okazaki K, Gardeniers JGE. Chemical Engineering Journal. 2011;**167**: 560

[71] Zhou J, Xu Y, Zhou X, Gong J, Yin Y, Zheng H, et al. Direct oxidation of methane to hydrogen peroxide and organic oxygenates in a double dielectric plasma reactor. ChemSusChem. 2011;**4**: 1095-1098

[72] Matin NS, Whitehead JC. A chemical model for the atmospheric pressure plasma reforming of methane with oxygen. In: 28th ICPIG, 15–20 July 2007; Prague, Czech Republic. 2007. p. 983

[73] Snoeckx R, Ozkan O, Reniers F, Bogaerts A. The quest for value-added products from carbon dioxide and water in a dielectric barrier discharge: A chemical kinetics study. ChemSusChem. 2017;**10**:409-424

[74] Eliasson B, Kogelschatz U, Xue B, Zhou LM. Hydrogenation of carbon dioxide to methanol with a discharge-activated catalyst. Industrial and Engineering Chemistry Research. 1998; **37**:3350-3357

[75] De Bie C, van Dijk J, Bogaerts A. CO$_2$ hydrogenation in a dielectric barrier discharge plasma revealed. Journal of Physical Chemistry C. 2016; **120**:25210-25224

[76] Heijkers S, Snoeckx R, Kozák T, Silva T, Godfroid T, Britun N, et al. CO$_2$ conversion in a microwave plasma reactor in the presence of N$_2$: Elucidating the role of vibrational levels. Journal of Physical Chemistry C. 2015; **119**:12815-12828

[77] Snoeckx R, Heijkers S, Van Wesenbeeck K, Lenaerts S, Bogaerts A. CO$_2$ conversion in a dielectric barrier discharge plasma: N$_2$ in the mix as a helping hand or problematic impurity? Energy & Environmental Science. 2016; **9**:999-1011

[78] Legrand J, Diamy A, Hrach R, Hrachova V. Kinetics of reactions in CH$_4$/N$_2$ afterglow plasma. Vacuum. 1997; **48**:671-675

[79] Majumdar A, Behnke JF, Hippler R, Matyash K, Schneider R. Chemical reaction studies in CH$_4$/Ar and CH$_4$/N$_2$ gas mixtures of a dielectric barrier discharge. The Journal of Physical Chemistry. A. 2005; **109**:9371-9377

[80] Pintassiglio CD, Jaoul C, Loureiro J, Belmonte T, Czerwiec T. Kinetic modelling of a N$_2$ flowing microwave discharge with CH$_4$ addition in the post-discharge for nitrocarburizing treatments. Journal of Physics D: Applied Physics. 2007; **40**:3620-3632

[81] Jauberteau JL, Jauberteau I, Cinelli MJ, Aubreton J. Reactivity of methane in a nitrogen discharge afterglow. New Journal of Physics. 2002; **4**:39

[82] Savinov SY, Lee H, Keun H, Na B. The effect of vibrational excitation of molecules on plasmachemical reactions involving methane and nitrogen. Plasma Chemistry and Plasma Processing. 2003; **23**:159-173

[83] Snoeckx R, Setareh M, Aerts R, Simon P, Maghari A, Bogaerts A. Influence of N$_2$ concentration in a CH$_4$/N$_2$ dielectric barrier discharge used for CH$_4$ conversion into H$_2$. International Journal of Hydrogen Energy. 2013; **38**: 16098-16120

[84] Bogaerts A, De Bie C, Snoeckx R, Kozák T. Plasma based CO$_2$ and CH$_4$ conversion: A modeling perspective. Plasma Processes and Polymers. 2017; **14**:e1600070

[85] Wang W, Snoeckx R, Zhang X, Cha MS, Bogaerts A. Modeling plasma-based CO$_2$ and CH$_4$ conversion in mixtures with N$_2$, O$_2$ and H$_2$O: The bigger plasma chemistry picture. Journal of Physical Chemistry C. 2018; **122**:8704-8723

[86] Wang W, Mei D, Tu X, Bogaerts A. Gliding arc plasma for CO$_2$ conversion: Better insights by a combined experimental and modelling approach. Chemical Engineering Journal. 2017; **330**:11-25

[87] Trenchev G, Kolev S, Wang W, Ramakers M, Bogaerts A. CO$_2$ conversion in a gliding arc plasmatron: Multidimensional modeling for improved efficiency. Journal of Physical Chemistry C. 2017; **121**:24470-24479

[88] Chang JS, Kostov KG, Urashima K, Yamamoto T, Okayasu Y, Kato T, et al. Removal of NF$_3$ from semiconductor-process flue gases by tandem packed-bed plasma and adsorbent hybrid systems. IEEE Transactions on Industry Applications. 2000; **36**:1251-1259

[89] Takaki K, Chang JS, Kostov KG. Atmospheric pressure of nitrogen plasmas in a ferroelectric packed bed barrier discharge reactor. Part I. modeling. IEEE Transactions on Dielectrics and Electrical Insulation. 2004; **11**:481-490

[90] Zhang Y, Wang HY, Jiang W, Bogaerts A. Two-dimensional particle-in cell/Monte Carlo simulations of a packed-bed dielectric barrier discharge in air at atmospheric pressure. New Journal of Physics. 2015; **17**:083056

[91] Kang WS, Park JM, Kim Y, Hong SH. Numerical study on influences of barrier arrangements on dielectric barrier discharge characteristics. IEEE Transactions on Plasma Science. 2003;**31**:504-510

[92] Russ H, Neiger M, Lang JE. Simulation of micro discharges for the optimization of energy requirements for removal of NO_x from exhaust gases. IEEE Transactions on Plasma Science. 1999;**27**:38-39

[93] Kruszelnicki J, Engeling KW, Foster JE, Xiong Z, Kushner MJ. Propagation of negative electrical discharges through 2-dimensional packed bed reactors. Journal of Physics D: Applied Physics. 2017;**50**:025203

[94] Van Laer K, Bogaerts A. Fluid modelling of a packed bed dielectric barrier discharge plasma reactor. Plasma Sources Science and Technology. 2016; **25**:015002

[95] Van Laer K, Bogaerts A. Influence of gap size and dielectric constant of the packing material on the plasma behaviour in a packed bed DBD reactor: A fluid modelling study. Plasma Processes and Polymers. 2017;**14**: e1600129

[96] Van Laer K, Bogaerts A. How bead size and dielectric constant affect the plasma behaviour in a packed bad plasma reactor: A modelling study. Plasma Sources Science and Technology. 2017;**26**:085007

[97] Wang W, Kim HH, Van Laer K, Bogaerts A. Streamer propagation in a packed bed plasma reactor for plasma catalysis applications. Chemical Engineering Journal. 2018;**334**: 2467-2479

[98] Kang WS, Kim HH, Teramoto Y, Ogata A, Lee JY, Kim DW, et al. Surface streamer propagations on an alumina bead: Experimental observation and numerical modeling. Plasma Sources Science and Technology. 2018;**27**: 015018

[99] Georgieva V, Berthelot A, Silva T, Kolev S, Graef W, Britun N, et al. Understanding microwave surface-wave sustained plasmas at intermediate pressure by 2D modeling and experiments. Plasma Processes and Polymers. 2017;**14**:e1600185

[100] Janssen GM. Design of a general plasma simulation model: Fundamental aspects and applications. PhD thesis. Eindhoven University of Technology, the Netherlands; 2000. 186 p

[101] Jimenez-Diaz M, Carbone EAD, van Dijk J, van der Mullen JJAM. A two-dimensional Plasimo multiphysics model for the plasma–electromagnetic interaction in surface wave discharges: The surfatron source. Journal of Physics D: Applied Physics. 2012;**45**:335204

[102] Rahimi S, Jimenez-Diaz M, Hubner S, Kemaneci EH, van der Mullen JJAM, van Dijk J. A two-dimensional modelling study of a coaxial plasma waveguide. Journal of Physics D: Applied Physics. 2014;**47**:125204

[103] Kabouzi Y, Graves DB, Castanos-Martınez E, Moisan M. Modeling of atmospheric-pressure plasma columns sustained by surface waves. Physical Review E. 2007;**75**:016402

[104] Richard F, Cormier JM, Pellerin S, Chapelle J. Physical study of a gliding arc discharge. Journal of Applied Physics. 1996;**79**:2245-2250

[105] Fridman A, Nester S, Kennedy LA, Saveliev A, Mutaf-Yardemci O. Gliding arc gas discharge. Progress in Energy and Combustion Science. 1999;**25**: 211-231

[106] Pellerin S, Richard F, Chapelle J, Cormier JM, Musiol K. Heat string model of bi-dimensional dc Glidarc. Journal of Physics D: Applied Physics. 2000;**33**:2407-2419

[107] Mutaf-Yardimci O, Saveliev AV, Fridman AA, Kennedy LA. Thermal and nonthermal regimes of gliding arc discharge in air flow. Journal of Applied Physics. 2000;**87**:1632-1641

[108] Kuznetsova IV, Kalashnikov NY, Gutsol AF, Fridman AA, Kennedy LA. Effect of "overshooting" in the transitional regimes of the low-current gliding arc discharge. Journal of Applied Physics. 2002;**92**:4231-4237

[109] Pellerin S, Cormier JM, Richard F, Musiol K, Chapelle J. Determination of the electrical parameters of a bi-dimensional d.c. Glidarc. Journal of Physics D: Applied Physics. 1999;**32**:891-897

[110] Gutsol AF, Gangoli SP. Transverse 2-D gliding arc modeling. IEEE Transactions on Plasma Science. 2017;**45**:555-564

[111] Kolev S, Bogaerts A. A 2D model for a gliding arc discharge. Plasma Sources Science and Technology. 2015;**24**:015025

[112] Kolev S, Bogaerts A. Similarities and differences between gliding glow and gliding arc discharges. Plasma Sources Science and Technology. 2015;**24**:065023

[113] Kolev S, Sun SR, Trenchev G, Wang W, Wang HX, Bogaerts A. Quasi-neutral modeling of gliding arc plasmas. Plasma Processes and Polymers. 2017;**14**:e1600110

[114] Sun SR, Kolev S, Wang HX, Bogaerts A. Coupled gas flow-plasma model for a gliding arc: Investigations of the back-breakdown phenomenon and its effect on the gliding arc characteristics. Plasma Sources Science and Technology. 2017;**26**:015003

[115] Trenchev G, Kolev S, Bogaerts A. A 3D model of a reverse vortex flow gliding arc reactor. Plasma Sources Science and Technology. 2016;**25**:035014

[116] Bogaerts A, Berthelot A, Heijkers S, Kolev S, Snoeckx R, Sun S, et al. CO₂ conversion by plasma technology: Insights from modeling the plasma chemistry and plasma reactor design. Plasma Sources Science and Technology. 2017;**26**:063001

[117] Fridman A. Plasma Chemistry. Cambridge: Cambridge University Press; 2008. 1022 p

[118] Hagelaar GJM, Pitchford LC. Solving the Boltzmann equation to obtain electron transport coefficients and rate coefficients for fluid models. Plasma Sources Science and Technology. 2005;**14**:722-733

[119] Bogaerts A, Kozák T, Van Laer K, Snoeckx R. Plasma-based conversion of CO₂: Current status and future challenges. Faraday Discussions. 2015;**183**:217-232

[120] Aerts R, Snoeckx R, Bogaerts A. In-situ chemical trapping of oxygen in the splitting of carbon dioxide by plasma. Plasma Processes and Polymers. 2014;**11**:985-992

[121] Kim HH, Wang N, Ogata A, Song YH. Propagation of surface streamers on the surface of HSY zeolites-supported silver nanoparticles. IEEE Transactions on Plasma Science. 2011;**39**:2220-2221

[122] Kim HH, Ogata A. Interaction of nonthermal plasma with catalyst for the air pollution control. International

journal of Environmental Science and Technology. 2012;**6**:43-48

[123] Tu X, Gallon HJ, Whitehead JC. Transition behavior of packed-bed dielectric barrier discharge in argon. IEEE Transactions on Plasma Science. 2011;**39**:2172-2173

Chapter 3

Plasma-Enabled Dry Methane Reforming

Zunrong Sheng, Seigo Kameshima, Kenta Sakata and Tomohiro Nozaki

Abstract

Plasma-enabled dry methane reforming is a promising technology for biogas upgrade and shows multiple benefits to provide additional energy and material conversion pathways. This chapter first presents the role of nonthermal plasma as a potential energy supply pathway in the low-temperature methane conversion: an appropriated combination of electrical energy provided by plasma (ΔG) and the low-temperature thermal energy ($T\Delta S$) satisfies the overall reaction enthalpy (ΔH) with higher energy conversion efficiency. Moreover, plasma-enabled dry methane reforming could be operated at much lower temperature than thermal catalysis with sufficient material conversion. Three kinds of typical packed-bed plasma reactor were introduced to give a better understanding of the application of plasma and catalyst hybrid system. Subsequently, plasma-enabled dry methane reforming was diagnosed by pulsed reaction spectrometry compared with thermal catalysis, presenting a clear overview of gas component changes and significant promotion in reactant conversion and product yield. The interaction between plasma and catalyst was summarized based on two aspects: catalyst affects plasma, and plasma affects catalyst. We discussed the coke formation behavior of Ni/Al_2O_3 catalyst in the plasma-enabled and thermal dry methane reforming, followed by the oxidation behavior. The interaction between plasma and catalyst pellets was discussed toward deeper insight into the mechanism.

Keywords: plasma catalysis, dry methane reforming, dielectric barrier discharge, biogas, methane conversion

1. Introduction

Dry methane reforming (DMR) has drawn keen attention as viable CO_2 utilization technology because it may have one of the greatest commercial potentials [1, 2].

$$CH_4 + CO_2 \rightarrow 2H_2 + 2CO \quad \Delta H = 247\,kJ/mol \quad \text{(R1)}$$

Moreover, products are the main components of syngas (H_2 and CO), which can be converted to the synthetic fuels as well as H_2 carrier via well-established C1 chemistry. Conventionally, the H_2/CO ratio from DMR is more suitable for Fischer-Tropsch synthesis than other methane reforming reactions [3–5]. **Figure 1** shows the reaction enthalpy and Gibbs free energy of DMR (R1) with respect to temperature. According to the definition, reaction enthalpy (Eq. (1)) consists of two terms:

$$\Delta H = T\Delta S + \Delta G \quad \text{(1)}$$

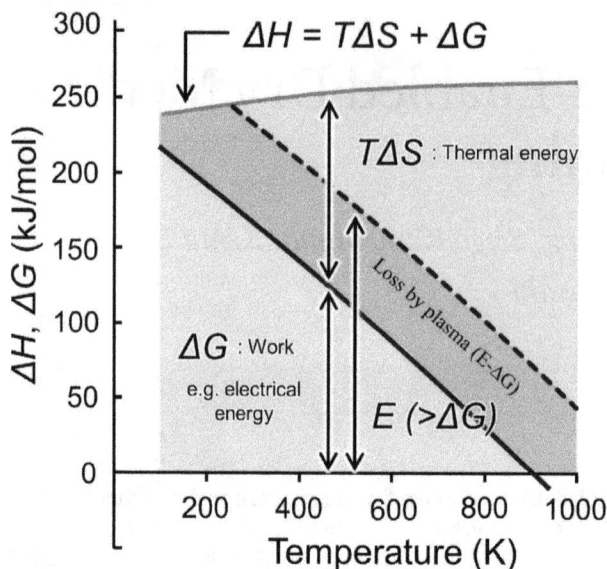

Figure 1.
Energy diagram of DMR.

DMR is categorized as uphill (endothermic) reaction where energy input (ΔH) is indispensable in order to satisfy the conservation of energy. Moreover, the reaction does not occur spontaneously by the low-temperature thermal energy due to the large positive value of ΔG at low temperature. **Figure 1** shows that at least 900 K is required to have a negative value of ΔG, and all energy is supplied via high-temperature thermal energy. Such high-temperature heat is supplied by the combustion of initial feed that produces CO_2 as well as NO_x. Net CO_2 utilization is partly canceled unless combustion-generated CO_2 is utilized which is economically quite difficult. Moreover, heat transfer from the combustion gas flowing outside of the reactor to the catalyst bed governs the overall material throughput which is known as a *heat transfer-limiting* regime. Because the heat transport property of a fixed bed reactor is poor, excessively high-temperature operation beyond thermodynamic limitation (i.e., 900 K) is necessary.

To overcome the aforementioned problem, low-temperature DMR is demanded, pursuing a new technology, and potential use of nonthermal plasma is highlighted. Assume DMR is operated at a lower temperature than the thermodynamic limitation as schematically depicted in **Figure 1**. A part of the energy is supplied by a low-temperature thermal energy ($T\Delta S$), while the rest of energy is supplied by the electricity (ΔG) under the nonthermal plasma environment so that $T\Delta S + \Delta G$ satisfies reaction enthalpy (ΔH). Electrical energy is used to accelerate electrons; subsequently, the electron energy is transferred to the molecules to initiate DMR at much lower temperature than thermal catalysis. Electronic collision process is independent of reaction temperature if gas density does not change significantly. Meanwhile, a part of the electrical energy is converted to heat: electrical energy consumed by nonthermal plasma (E) is depicted in the dashed line in **Figure 1**: inevitably, E is greater than ΔG at a fixed temperature. Although heat generated by nonthermal plasma is considered as energy loss (i.e., $E-\Delta G$), both excited species and heat are utilized via endothermic DMR, which enables efficient use of electrical energy without *heat transfer limitation*: electrification of reforming reaction, or chemical processes in general, has the greatest advantage that the energy transfer and the control are independent of temperature gradient.

Dielectric barrier discharge (DBD) is the most successful atmospheric pressure nonthermal plasma sources in industry applications [6] and is used exclusively for this purpose. DBD is combined with a catalyst bed reactor and generated at atmospheric pressure [7]. DBD is characterized as a number of transient discharge channels known as streamers with nanosecond duration. Because the streamer has a nature of propagation along the interface between two adjacent dielectric materials, namely, the catalyst pellet and the gas interface, excited species produced by DBD is transferred to the catalyst surface efficiently. Moreover, the heat generated by DBD is transferred directly to the catalysts; overall energy transfer from nonthermal plasma to the catalyst bed is efficient. If the electricity is supplied from the renewable energy such as photovoltaics and wind turbines, low-emission DMR is possible with free of combustion. Moreover, nonthermal plasma-assisted C1 chemistry enables renewable-to-chemical energy conversion, which provides an alternative and viable solution for the efficient renewable energy storage and transportation pathways.

The aforementioned thermodynamic analysis (**Figure 1**) implies that the temperature-benign and low-emission chemical processes are possible with the appropriate combination of nonthermal plasma and the heterogeneous catalysts. Meanwhile, such hybrid system does not work at room temperature simply because the overall reaction rate is *kinetically controlled* at much low temperature: Nevertheless, we would like to highlight that nonthermal plasma technology solves many technological obstacles such as the elimination of combustion as well as heat transfer limitation. Moreover, low-temperature operation suppresses coke formation which is one of the big issues in DMR. In this book chapter, we focus on low-temperature DMR and compare thermal and plasma catalysis. Plasma catalysis of DMR was diagnosed by pulsed reaction spectrometry [8], and results were compared with thermal catalysis to highlight the benefit of DBD and catalyst combination. Subsequently, the interaction between DBD and catalyst pellets was discussed toward deeper insight into the mechanism. Finally, future prospects of plasma catalysis of DMR are provided.

2. Packed-bed plasma reactor

Based on the location and number of plasma zone and catalyst bed, the combination of heterogeneous catalysts with plasma can be operated in three configurations: single-, two- and multistage, which are shown in **Figure 2**.

2.1 Single-stage reactor

In a single-stage reactor (**Figure 2(a)**), the catalyst is packed inside the plasma zone, where the interaction of plasma and catalysts occurs. Because thermal plasma (gas bulk temperature >3000°C [9]) could damage catalyst, single-stage reactor is, therefore, suitable for nonthermal plasma sources. The single-stage reactor is widely applied in CH_4 reforming [10–14], direct conversion of CO_2 [15–19], VOCs abatement [20–22], exhaust matter removal [23], formaldehyde removal [24], NO_x synthesis [25], and ozone synthesis [6]. There are two significant merits of single-state reactor: (I) great flexibility exists in terms of electrode and reactor configurations that the reactor can be constructed using inexpensive materials such as glass and polymers and (II) reactive species, ions, electrons, etc. generated by nonthermal plasma could modify the gas composition, which affects the surface reactions with catalyst synergistically. However, the interaction between plasma and catalyst is complex when the catalyst is placed directly in the plasma zone. The synergy of plasma and catalyst will be discussed further in Section 4 based on the single-state reactor.

(a) **Singe-stage reactor**

(b) **Two-stage reactor**

CAT-A   CAT-B   CAT-C

(c) **Multistage reactor**

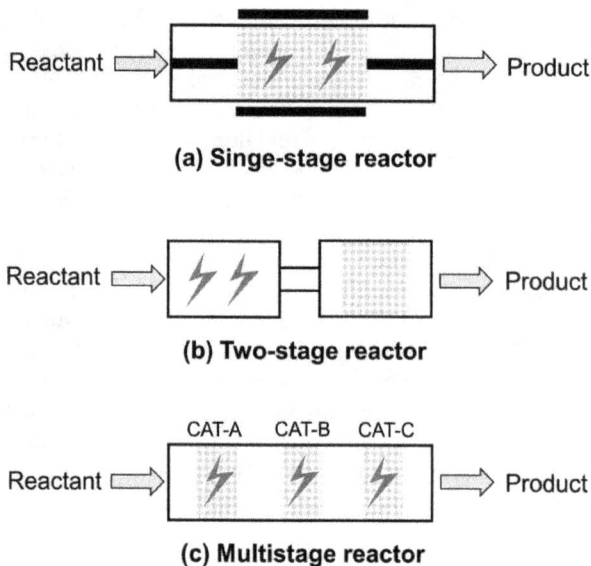

Figure 2.
Schematic diagram of single-stage (a), two-stage (b), and multistage reactor (c). Catalyst is depicted as orange circle; plasma is depicted as purple "lightning" symbol.

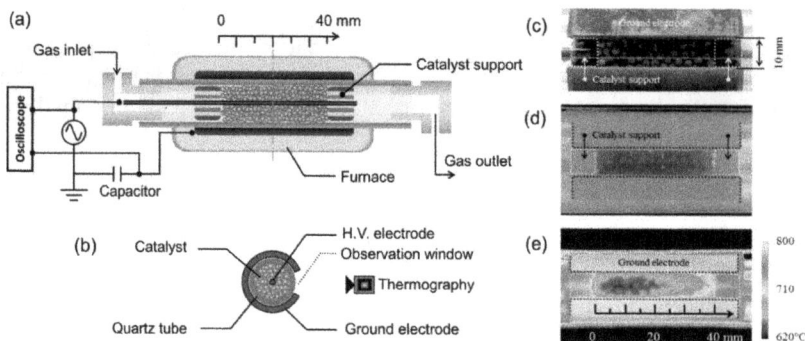

Figure 3.
Single-stage DBD reactor system for DMR: (a) overview of the reactor system, (b) cross-sectional view, (c) overview of the catalyst bed, (d) DBD generated in the catalyst bed, and (e) temperature distribution during reforming reaction.

Figure 3 depicts a single-stage DBD reactor for CH_4 reforming [26], mainly including a quartz tube reactor, high voltage (HV) electrode at the center, and ground electrode outside of the tube. Catalyst pellets are packed in the plasma zone between two electrodes, and both ends of the catalyst bed are fixed by metallic supports. The high voltage is applied between the HV centered electrode and ground electrode to generate dielectric barrier discharge over the pellet surface. Discharge power was measured by voltage-charge Lissajous analysis. The discharge gap, which is the distance between HV electrode and the ground electrode, is usually less than 10 mm [27]. Catalyst temperature is controlled by a furnace. The temperature distribution of the catalyst bed is measured by thermography through the observation window. **Figure 3(c)–(e)** shows an overview of the catalyst bed, DBD generated in the catalyst bed, and the temperature distribution during reforming reaction. The catalyst bed temperature was clearly decreased because of the endothermic nature of DMR. In addition, gas temperature was estimated by optical emission spectroscopy (OES) of CO(B-A) transition [28], showing that catalyst temperature and gas temperature matched within a measurement error.

2.2 Two-stage reactor

In the two-stage reactor, the catalyst is located at the downstream of plasma (**Figure 2(b)**). The gas is first addressed by the plasma and subsequently interacts with the catalyst [29]. Due to the separation of plasma and catalyst, both thermal and nonthermal plasma can be utilized. Because excited species generated in plasma have very short lifetimes, plasma mainly plays the role to preconvert the gas composition and then feed it into the catalyst reactor, e.g., in NO_x removal process, due to the pretreatment of plasma, NO and NO_2 were coexisted, which enhanced the following selective catalytic reduction in catalyst bed [30]; the other example is the benzene removal process where ozone (O_3) was formed from background O_2 by plasma, which promoted the decomposition of benzene in the next stage [31]. However, compared with the single-stage reactor, application of a two-stage reactor is limited in plasma catalysis and shows a lower performance for a given catalyst [32–37].

2.3 Multistage reactor

The multistage reactor can be described as a combination of more than one single-stage bed/reactor (**Figure 2(c)**). The multistage reactor gives a more flexible option in the industrialization of the plasma catalysis, attributing to the combination of catalysts with a different function for the expected reaction [38]. Chavade et al. [39] used a four-stage plasma and catalytic reactor system for oxidation of benzene. The results showed that the increase in stage number enhanced benzene conversion and CO_2 selectivity. The same result can be found in biogas reforming process using a multistage gliding arc discharge system without catalyst [40]. Harling et al. [41] developed a three-stage reactor for VOCs removal. The combination of plasma and catalyst in series could significantly improve the efficiency of VOCs decomposition. At the same time, the formation of by-product such as NO_x was suppressed.

3. DBD-enabled dry methane reforming

The pulsed reaction spectrometry using DBD with Ni/Al_2O_3 catalysts was investigated to develop a reforming diagnostic method [10]. Pulsed reforming enables the transient analyses of both CH_4/CO_2 consumption and H_2 and CO generation. Furthermore, carbon formation was analyzed quantitatively without serious catalyst deactivation. The varied CH_4/CO_2 ratios between 0.5 and 1.5 were investigated at a fixed catalyst temperature near 600°C. The CH_4/CO_2 ratio was initially set to 0.5, and then the CH_4/CO_2 ratio was incremented stepwise until $CH_4/CO_2 = 1.5$, consecutively, while total flow rate was fixed at 1000 cm^3/min. De-coking process (R2) was followed up after every pulsed reaction. System pressure was kept at 5 kPa during the reforming process. Discharge power was 85–93 W where specific energy input was ca. 1.2 eV/molecule. Commercially available catalyst pellets (11 wt% $Ni-La/Al_2O_3$, Raschig ring type: 3 mm) was packed for 40 mm length (total weight ca. 12 g; Ni 1.36 g; La 0.35 g). **Figure 4** provides an overview of gas component changes in the entire hybrid reforming.

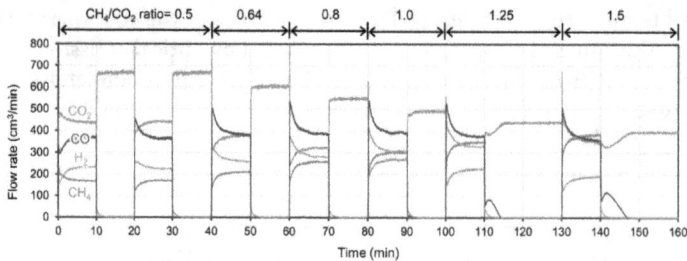

Figure 4.
Overview of the entire pulsed hybrid DMR.

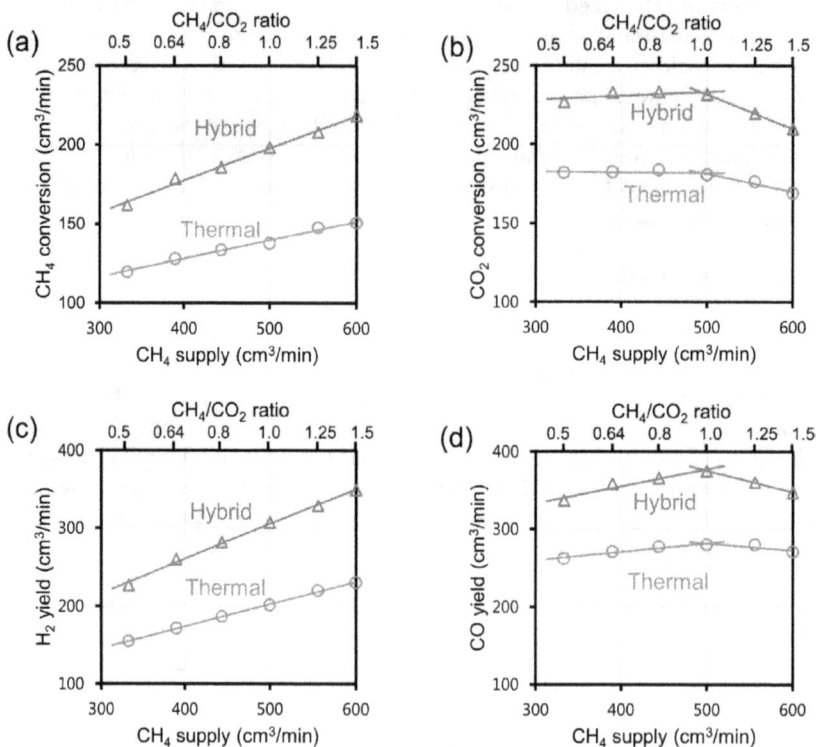

Figure 5.
Effects of CH_4/CO_2 ratio on DMR at ca. 600°C: (a) CH_4 conversion, (b) CO_2 conversion, (c) H_2 yield, and (d) CO yield.

Reactant conversion and product yields are shown in **Figure 5**. The definition for conversion and yield were provided in Ref. [8]. CH_4 conversion and H_2 yield were monotonically increased with the CH_4/CO_2 ratio. There are two simultaneous routes for CH_4 conversion as shown in **Figure 6**. Route (I) is a reforming path: CH_4 is chemisorbed on metallic sites (adsorbed species are denoted by * in reaction). The adsorbed CH_4 fragments (CH_x*) is oxidized by CO_2* to form CH_xO* before complete dehydrogenations to C* occurs. In route (II), CH_4 almost irreversibly dehydrogenates toward carbon atom, and then C*-rich layer is oxidized slowly by CO_2* (R2), which can be evidenced in the de-coking process in **Figure 4**. When the CH_4/CO_2 ratio exceeded 1.0, CH_4 prefers to dehydrogenate to solid carbon through route (II) due to the low proportion of CO_2. Subsequently, a nonnegligible amount of solid carbon is produced, and CO_2 conversion and CO yield turned to proportionally decrease.

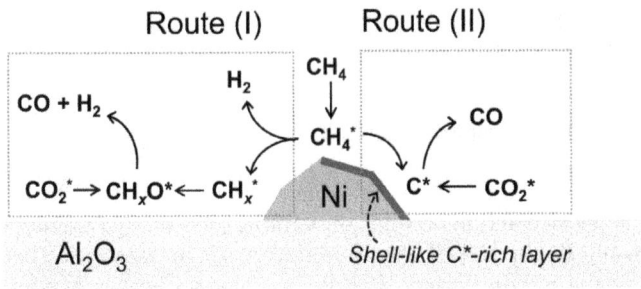

Figure 6.
Two simultaneous routes for CH$_4$ conversion.

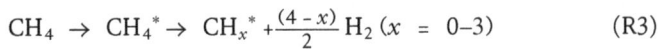

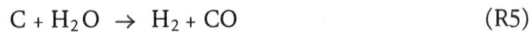

$$C + CO_2 \rightarrow CO + CO \qquad\qquad (R2)$$

$$CH_4 \rightarrow CH_4^* \rightarrow CH_x^* + \frac{(4-x)}{2} H_2 \ (x = 0\text{--}3) \qquad\qquad (R3)$$

$$H_2 + CO_2 \rightarrow CO + H_2O \qquad\qquad (R4)$$

$$C + H_2O \rightarrow H_2 + CO \qquad\qquad (R5)$$

Compared with thermal reforming, both CH$_4$ conversion and H$_2$ yield were clearly promoted in hybrid reforming (**Figure 5**), and the main pathway of CH$_4$ conversion and H$_2$ yield could be simply described as CH$_4$ dehydrogenation (R3). It is proposed that CH$_4$ dehydrogenation was enhanced by the synergistic effect of DBD and catalyst. Molecular beam study revealed that dissociative chemisorption of CH$_4$ on the metal surface was enhanced by vibrational excitation [42]. The numerical simulation of one-dimensional streamer propagation demonstrated that the vibrationally excited CH$_4$ is the most abundant and long-lived species generated by low-energy electron impact [43]. The reaction mechanism of plasma-enabled catalysis could be explained by the Langmuir-Hinshelwood (LH) reaction scheme. The analysis of overall activation energy is expected to understand the contribution of plasma-generated reactive species.

The CO$_2$ conversion and CO yield were promoted in hybrid reforming compared to thermal reforming (**Figure 5**). H$_2$O was simultaneously produced as a by-product by reverse water gas shift (RWGS) reaction (R4). Reactivity of plasma-activated H$_2$O was confirmed by Arrhenius plot analysis where reaction order for H$_2$O was doubled by DBD [44]. Plasma-activated H$_2$O promotes reaction with adsorbed carbon; it creates additional pathways (R5) to syngas (H$_2$ and CO). The CO$_2$ conversion and CO yield were promoted in the hybrid reforming, illustrating that the reverse-Boudouard reaction (R2) was enhanced by DBD. The reaction between plasma-activated CO$_2$ and adsorbed carbon increases CO yield. The same result was obtained in the de-coking period [10]. Although excessive production of carbon is detrimental for catalyst activity and lifetime, the presence of adsorbed carbon creates key pathways for emerging plasma-induced synergistic effect. Consequently, plasma-activated CO$_2$ and H$_2$O would promote surface reaction and increase CO and H$_2$ yield. **Figure 5** clearly shows that the slope of each line increased in hybrid reforming compared with thermal reforming, attributing to the nonthermal plasma-excited species. The increase of slope could be further explained by the promoted overall reaction order, which plays the key role in the estimation of the rate-determining step [44].

4. The synergy of plasma and catalyst

Synergism in plasma catalysis in the single-stage reactor is not fully understood due to the complex interaction between the various plasma-catalyst interaction processes [45–49]. Kim et al. [27] discussed the criteria for interaction between nonthermal plasma and the porous catalyst. The chemical species in nonthermal plasma are highly reactive; the lifetime is very short, e.g., O (^{1}D), 10 ns; O (3P), 50 µs; and OH, 100 µs: their one-dimensional diffusion length is limited from 0.7 to 65 µm. Plasma generated species within diffusion length from the external surface of pellets would contribute to the plasma-induced reaction pathways. The interdependence of plasma and catalyst can be discussed as two aspects: catalyst affects plasma, and plasma affects catalyst.

4.1 The effects of catalyst on plasma

With the packed catalyst in the plasma zone, gaseous species adsorbed on the catalyst surface increase the concentration of surface species. In addition, the electric field is enhanced near the catalyst surface due to catalyst nanofeatures [50, 51]. Moreover, the packed catalyst also enables the discharge type change, as well as microdischarge generation.

Without packed materials, discharge mode is the "free-standing" filamentary discharge propagating across the gas gap (**Figure 7(a)**). With a packed catalyst,

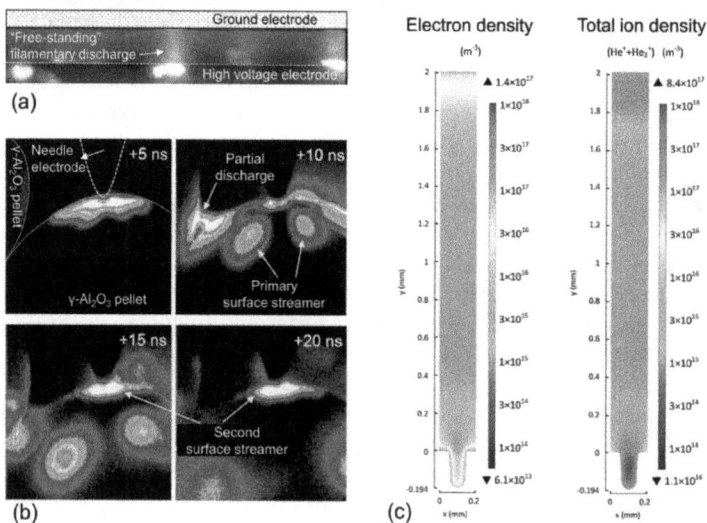

Figure 7.
(a) Filamentary discharge without catalyst pellet, (b) time-resolved images of surface streamers propagating on the surface of γ-Al₂O₃ (reproduced with permission from ref 54. Copyright 2016 IOP Publishing), and (c) distributions of the electron density and total ion density with a 100 µm pore (reproduced with permission from ref 56. Copyright 2016 Elsevier).

the surface streamer is propagated with the close contact with the catalyst surface, and intensive partial discharges occur between the contact area of catalysts [52, 53]. The time-resolved visualization of surface streamer propagation and partial discharge were detected by Kim et al. [54] with an intensified charge-coupled device (ICCD) camera. **Figure 7(b)** shows time-resolved images of surface streamers (i.e., primary surface streamer and secondary surface streamer) propagating on the surface. Enhanced catalytic performance in the presence of a catalyst is closely linked with the propagation of surface streamers [55]. Moreover, with packed catalyst, microdischarge is generated inside catalyst pores (when the pore sizes >10 μm) [56, 57]. Zhang et al. [56] investigated microdischarge formation inside catalyst pores by a two-dimensional fluid model in the μm range (**Figure 7(c)**), indicating that the plasma species can be formed inside pores of structured catalysts in the μm range and affect the plasma catalytic process.

4.2 The effects of plasma on catalyst

The plasma also affects the catalyst properties, which are summarized as the following aspects:

(I) Modification of physicochemical properties of the catalyst by plasma, which is widely used in catalyst preparation processes [58]. With plasma preparation, the catalyst obtains a higher adsorption capacity [59], higher surface area, and higher dispersion of the catalyst material [60–62], leading to a plasma-enhanced reactivity.

(II) It is possible that plasma makes changes in the surface process with the catalyst. As for the CH_4 reforming process, the deposited carbon from Ni catalysts can be removed effectively by plasma-excited CO_2 and H_2O [8]. The other example of the synergism is NH_3 decomposition for the application of fuel cell, where NH_3 conversion reached 99.9% when combining plasma and catalyst, although the conversion was less than 10% in the case of either plasma or catalyst only reaction [63].

(III) Based on Arrhenius plot analysis, plasma can decrease the activation barriers (overall activation energy), attributed to the vibrational excitation, which is schematically depicted in **Figure 8**. The net activation barrier will be $\left(E_a^{v=0} - E_{vib}\right)$ in an adiabatic barrier crossing case and $\left(E_a^{v=0} - E_{vib} - E^*\right)$ in a nonadiabatic barrier crossing case, respectively [64]. The activation barrier decrease was reported in toluene destruction process [65]. In steam methane reforming, the preexponential factor was increased clearly by plasma, attributing to plasma-activated H_2O removes adsorbed carbon species, which regenerate active sites for subsequent CH_4 adsorption [11].

(IV) The excited species or dissociated species might create other pathways with the presence of catalyst, e.g., during the CO_2 plasma oxidation process, plasma-enhanced CO_2 oxidized Ni/Al_2O_3 catalyst to form a NiO layer, which could drive an oxidation–reduction cycle in dry methane reforming reaction. The same NiO layer was found when specific energy input (*SEI*) was sufficiently high: the O_2 that dissociated from CO_2 plays the key role in the oxidation process. The details will be interpreted in Section 5.2.

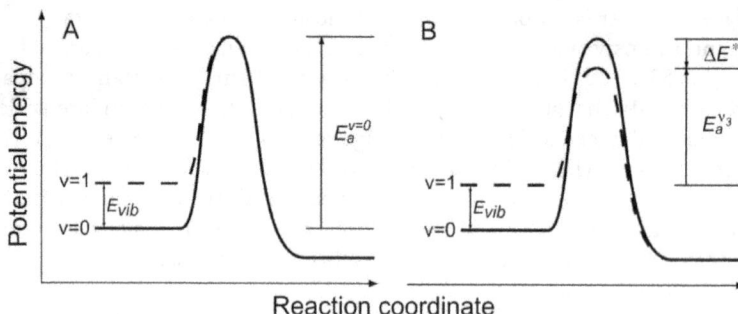

Figure 8.
Reduction of the overall activation energy by vibrational excitation of the reactants. (A) Adiabatic barrier crossing case and (B) nonadiabatic barrier crossing case. Reprinted with permission from ref 64. Copyright 2004 AAAS.

5. Discussion

5.1 Coke formation behavior

Coke formation behavior was studied as a reaction footprint to track reaction pathways induced by DBD [26]. Coke morphology and their distribution over the 3 mm spherical Ni/Al$_2$O$_3$ catalyst pellets were obtained after 60 min DMR. **Figure 9** shows cross-sectional carbon distribution, where (a)–(c) and (d)–(f) correspond to plasma catalysis and thermal catalysis in low, middle, and high temperatures. For the thermal catalysis with the temperature at 465°C, carbon deposition over the entire cross-section was obvious. With the temperature increased, coke was decreased and finally became nondetectable at ca. 620°C. At low temperature, plasma catalysis suppressed the coke formation significantly over the entire cross-section.

By the analysis of scanning electron micrographs (SEM) and Raman spectrum, fine carbon filaments were detected on the external pellet surface in plasma catalysis [26]. In contrast, thick fibrous carbon deposition was observed on the external surface in thermal catalysis, as well as in the internal pores in both thermal and plasma

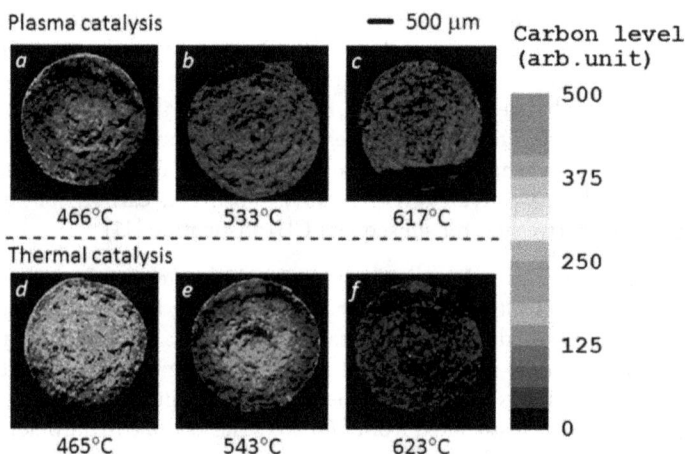

Figure 9.
Carbon distribution over the 3 mm spherical pellet cross-section after 60 min reforming: plasma catalysis (a)–(c) and thermal catalysis (d)–(f), respectively. Reprinted with permission from ref 26. Copyright 2018 IOP Publishing.

catalyses. The CH_4 dehydrogenation on catalyst is enhanced by nonthermal plasma, contributing to the generation of highly filamentous and amorphous carbon. Such nonthermal plasma-enhanced reaction has been demonstrated by carbon nanotube growth [66] and plasma-enabled steam methane reforming [67]. The fine amorphous carbon filaments, deposited in the external surface of catalyst, prove that the interaction of DBD occurs mainly in the external surface. Consequently, DBD generation and plasma-excited species diffusion are inhibited in the internal pores of catalyst.

5.2 Oxidation behavior of Ni/Al$_2$O$_3$ catalyst

The nickel (Ni) of Ni/Al$_2$O$_3$ catalyst was oxidized slightly by CO_2 in the thermal catalysis [68, 69] . However, the significant Ni oxidation by CO_2 (R6) was demonstrated when the catalyst were packed in nonthermal plasma zone. In this case, Ni uptakes surface oxygen beyond the adsorption/desorption equilibrium (i.e., Langmuir isotherm) to form NiO, which further promotes CH_4 dehydrogenation without solid carbon deposition (R7):

$$Ni + CO_2 \rightarrow NiO + CO \tag{R6}$$

$$NiO + CH_4 \rightarrow Ni + OH + CH_3 \tag{R7}$$

The specific energy input (*SEI*) is a critical operational parameter in plasma-enabled CO_2 treatment process due to the fact that dominant reaction pathway shifts dramatically with *SEI*:

$$SEI = C \times \frac{Discharge\ power\ \text{(W)}}{Total\ flow\ rate\ \text{(cm}^3\text{/min)}} \text{ (eV/molecule)} \tag{2}$$

SEI expresses energy consumption by discharging per unit volume of the feed gas, which could be further interpreted as average electrical energy (eV) per molecule. In Eq. (2), C is the conversion factor of the unit [10]. Two contrasting conditions are demonstrated in plasma-enabled CO_2 oxidation: one is designated as the direct oxidation route; with a small *SEI*, CO_2 dissociation to CO and $0.5\ O_2$ (R9) is negligible, and then the plasma-excited CO_2 dominates the oxidation process (R8). The other is the indirect oxidation route where O_2 provides an additional oxidation pathway; with a large *SEI*, CO_2 is dissociated into CO and O_2 (R9) without heterogeneous catalysts by electron impact [70, 71], followed by Ni oxidation by O_2 (R10).

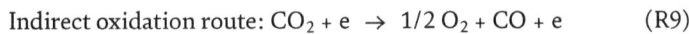

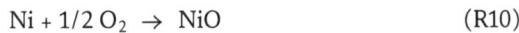

$$\text{Direct oxidation route: } Ni + CO_2 \rightarrow NiO + CO \tag{R8}$$

$$\text{Indirect oxidation route: } CO_2 + e \rightarrow 1/2\ O_2 + CO + e \tag{R9}$$

$$Ni + 1/2\ O_2 \rightarrow NiO \tag{R10}$$

The plasma-enhanced direct oxidation route (R8) is further investigated because the plasma-enabled synergistic effect was demonstrated distinctly without O_2 [8, 10, 26]. Ni oxidation behavior without O_2 was studied with *SEI* = 0.46 eV/molecule. The CO_2 conversion is far below 1% when the *SEI* was smaller than 0.5 eV/molecule [72, 73]; in the plasma and thermal oxidation, the CO_2 flow rate, catalyst temperature, and the oxidation time were controlled as 1000 cm^3/min, 600°C, and 70 min, respectively.

After DBD-enhanced oxidation and thermal oxidation, the formation of NiO and its distribution over the cross-section of 3 mm spherical pellet were investigated by Raman spectroscopy and optical microscope. Results showed that the NiO layer was recognized clearly with the thickness of ca. 20 μm. In contrast, the NiO layer was not identified after thermal oxidation. We should point out that the plasma-excited CO_2 has a strong oxidation capability of Ni catalyst. In addition, the effect of DBD is inhibited in the internal pores beyond 20 μm from the pellet surface.

In the thermal oxidation, CO_2 is most likely adsorbed at the perimeter between Ni nanocrystals and Al_2O_3 interfaces [74–76] (**Figure 10(a)**). Subsequently, the adsorbed CO_2 oxidize Ni to NiO near the perimeter. It is clear that the reaction sites for thermal oxidation are limited in the perimeter. The Ni oxidation reaction terminates after the reaction sites are fully oxidized by adsorbed CO_2. In plasma-enhanced oxidation reaction, CO_2 is firstly excited by electron impact. The vibrationally excited CO_2 plays the key role to enhance adsorption process and subsequent oxidation reaction of Ni catalyst, leading to an extensive Ni nanoparticle oxidation, which occurs not only in the perimeter but also in the terrace, step, and kink (**Figure 10(b)**). **Figure 10(c)** and (**d**) show hemispherical catalyst pellets after thermal and plasma oxidation. After thermal oxidation, the external surface and cross-section of catalyst pellets remained black. In contrast, after plasma oxidation, the external surface was oxidized and has showed whitish color change (oxidized stage); in the meantime, the cross-section of the hemispherical pellet has been kept black (unoxidized stage).

The vibrationally excited CO_2 by DBD would induce Ni oxidation to form the oxygen-containing active species (i.e., NiO) rather than simple adsorption, leading to oxygen-rich surface beyond Langmuir isotherm. Incoming plasma-excited CO_2 would carry a few eV internal energy due to the gas phase vibration-to-vibration energy transfer [77, 78], which is the main source of energy for NiO formation. Plasma-excited CO_2 could promote the adsorption flux; however, the adsorbed CO_2 is finally desorbed by the equilibrium limitation unless it forms NiO. In addition, the plasma-induced nonthermal heating mechanism plays another key role in the

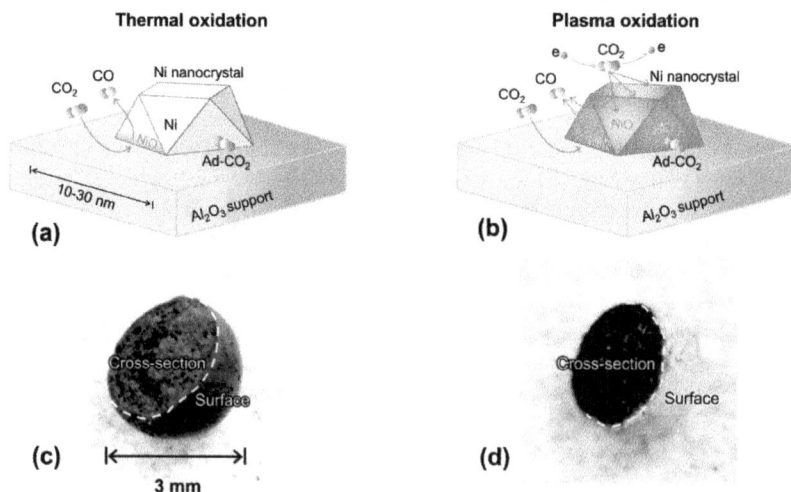

Figure 10.
Ni oxidation pathways: (a) thermal oxidation including CO_2 adsorption near the perimeter of Ni catalysts and (b) plasma-enhanced oxidation. Hemispherical catalyst pellets: (c) thermal oxidation and (d) plasma oxidation.

enhancement of Ni oxidation. Charge recombination and association of radicals can release energy corresponding to 1–10 eV/molecule on catalyst surface. When this excess energy is directly transferred to the adsorbed CO_2, Ni oxidation may be enhanced without increasing macroscopic catalyst temperature. This reaction scheme is explained by nonthermal plasma-mediated Eley-Rideal mechanism, rather than precursor-type adsorption enhancement.

In the DMR process, the oxygen-rich surface (NiO layer) has a capability of oxidizing a large flux of ground-state CH_4 efficiently. Consequently, CH_4 is not necessarily preexcited. CH_4 is almost fully reacted in the NiO layer (20 μm thickness) to inhibit the coke deposition toward the internal pores [26]. However, in the thermal catalysis, NiO is generated in a negligible amount. The ground-state CH_4 can diffuse into internal pores and deposit coke as previously confirmed [26]. Formation of NiO shell (**Figure 10(d)**) and the coke formation behavior (**Figure 9**) are well correlated in plasma catalysis as further discussed in the next section.

5.3 Interaction between DBD and catalyst pores

Although the synergies of plasma and catalyst have been summarized in Section 4, the interaction between DBD and catalyst pores will be further discussed in this section based on the carbon formation and oxidation behavior, as well as DBD-enhanced DMR. For the plasma catalysis, carbon deposition in the internal pores could be remarkably prevented, and fine amorphous carbon filaments were deposited only on the external surface of pellets. A similar trend was observed when NiO was formed in the limited region over the external surface (20 μm depth) only when DBD was superimposed. The results of coke formation behavior and oxidation behavior of Ni-based catalyst in plasma catalysis evidence that the interaction of DBD and catalyst occurs at the external surface of the pellets and the effected thickness is ca. 20 μm. Neither generation of DBD nor diffusion of plasma-generated reactive species in the internal pores is possible. Although DBD and pellet interaction is limited in the external surface, conversion of CH_4 and CO_2 was promoted clearly compared with thermal catalysis: this is the clear evidence of reaction enhancement by DBD.

For DBD, due to the enhanced physical interaction between propagating streamers and catalysts, plasma and catalyst contact area, as well as the streamer propagation from one pellet to the other, are promoted significantly. Nevertheless, electron density in a narrow filamentary channel is of the order of 10^{14} cm^{-3} [6, 79, 80]; in contrast, molecule density at standard condition is approximately 10^{19} cm^{-3}, indicating that a major part of the gas stream is neither ionized nor excited. Consequently, the extremely low proportion of ionized and excited species is inadequate to explain the net increase of CH_4 and CO_2 conversion and selectivity change by DBD. However, if reactive species are fixed and accumulated on the surface of the catalyst, the gross conversion of materials will be promoted. For this reason, the hetero-phase interface between DBD and the external pellet surface provides the most important reaction sites. In Section 5.2, nonthermal plasma oxidation of Ni to NiO creates a critically important step for plasma-enabled synergistic effect.

As Section 4.1 mentioned, gas breakdown is hard to occur in a pore smaller than 10 μm. For the pores catalyst with a pore size less than 2 nm, standard Paschen-type gas breakdown is impossible. To sum up, the external surface of pellet plays the key role for the DBD and catalyst interaction; however, the internal pores play a minor role.

6. Conclusion and outlook

The synergistic effect induced by DBD was clearly observed both in the CH_4 and CO_2 conversion and in the syngas yield. CH_4 dehydrogenation was enhanced by the synergistic effect of DBD and catalyst. Plasma-activated CO_2 and H_2O would promote surface reaction and increase CO and H_2 yield. The analysis of overall activation energy is expected to understand the contribution of plasma-generated reactive species.

In plasma catalysis, the fine amorphous carbon filaments, deposited in the external surface of catalyst, prove that the interaction of DBD occurs mainly in the external surface. The DBD generation and plasma-excited species diffusion are inhibited in the internal pores of the catalyst. Moreover, although the interaction between plasma and catalyst is limited in the external surface, the coke deposition was inhibited significantly in the internal pores by DBD, which is the clear evidence of reaction enhancement by DBD.

Oxidation behavior of Ni-based catalyst in nonthermal plasma-enabled catalysis showed that the NiO layer was generated in the external surface with the thickness of ca. 20 µm during plasma oxidation. In the internal pores, Ni oxidation is inhibited due to the negligible interaction with DBD. Contributing to the NiO layer, the surface of catalyst uptakes more oxygen beyond thermal equilibrium, which is known as Langmuir isotherm, creating a new reaction pathway via NiO. In the plasma catalysis of DMR, NiO drives the oxidation-reduction cycle, which promotes CH_4 dehydrogenation on the surface. Consequently, carbon deposition is suppressed effectively.

For further improvement of plasma-enhanced DMR, the following issues should be investigated:

(1) The effect of radical injection on reaction enhancement should be kinetically analyzed by the Arrhenius plot method, and the analysis of the overall activation energy is expected to understand the contribution of plasma-generated reactive species.

(2) Exploring new types of catalysts, dedicated to plasma catalysis, is an important subject of research. We have demonstrated that the interaction of DBD and catalyst occurs only at the external surface of the pellets, and the effected thickness is ca. 20 µm, which means a majority of the active sites in pores of catalyst do not interact with any excited species. New catalyst preparation method such as catalytic functionalization of reactor wall and catalyst coating for the reactor may be beneficial to strengthen the synergistic effect of nonthermal plasma and catalytically functionalized surface.

(3) The catalyst activity of partially oxidized catalyst and the nonthermal plasma heating mechanism have not been demonstrated experimentally yet; moreover, diagnosis of intermediate species on the surface, created by plasma-derived species, as well as their reaction dynamics are expected to be investigated for deep insight in plasma catalysis.

(4) Although the plasma-induced energy transfer mechanism is commonly accepted in particle growth, it has yet to be investigated within the scope of plasma catalysis. Deep understanding of highly transient and nonequilibrium energy transfer via excited molecules, without macroscopic temperature change, need to be studied.

(5) The individual contribution of radical injection and heat generation, as well as combination of those, must be understood. The gap between macroscopic and microscopic understanding, including various time scales covering nanoseconds to the millisecond, should be bridged by consistent manner.

Author details

Zunrong Sheng, Seigo Kameshima, Kenta Sakata and Tomohiro Nozaki*
Department of Mechanical Engineering, Tokyo Institute of Technology, Tokyo,
Japan

*Address all correspondence to: nozaki.t.ab@m.titech.ac.jp

IntechOpen

References

[1] Ravanchi MT, Sahebdelfar S. Carbon dioxide capture and utilization in petrochemical industry: Potentials and challenges. Applied Petrochemical Research. 2014;**4**:63-77. DOI: 10.1007/s13203-014-0050-5

[2] Shah YT, Gardner TH. Dry reforming of hydrocarbon feedstocks. Catalysis Reviews. 2014;**56**:476-536. DOI: 10.1080/01614940.2014.946848

[3] Fan MS, Abdullah AZ, Bhatia S. Catalytic technology for carbon dioxide reforming of methane to synthesis gas. ChemCatChem. 2009;**1**:192-208. DOI: 10.1002/cctc.200900025

[4] York APE, Xiao TC, Green MLH, Claridge JB. Methane oxyforming for synthesis gas production. Catalysis Reviews. 2007;**49**:511-560. DOI: 10.1080/01614940701583315

[5] Rostrup-Nielsen JR. New aspects of syngas production and use. Catalysis Today. 2000;**63**:159-164. DOI: 10.1016/S0920-5861(00)00455-7

[6] Kogelschatz U. Dielectric-barrier discharges: Their history, discharge physics, and industrial applications. Plasma Chemistry and Plasma Processing. 2003;**23**(1):1-46. DOI: 10.1023/A:1022470901385

[7] Chang MB, Yu SJ. An atmospheric-pressure plasma process for C_2F_6 removal. Environmental Science & Technology. 2001;**35**:1587-1592. DOI: 10.1021/es001556p

[8] Kameshima S, Tamura K, Mizukami R, Yamazaki T, Nozaki T. Parametric analysis of plasma-assisted pulsed dry methane reforming over Ni/Al$_2$O$_3$ catalyst. Plasma Processes and Polymers. 2017;**14**:1600096. DOI: 10.1002/ppap.201600096

[9] Schutze A, Jeong JY, Babayan SE, Jaeyoung P, Selwyn GS, Hicks RF. The atmospheric-pressure plasma jet: A review and comparison to other plasma sources. IEEE Transactions on Plasma Science. 1998;**26**:1685-1694. DOI: 10.1109/27.747887

[10] Kameshima S, Tamura K, Ishibashi Y, Nozaki T. Pulsed dry methane reforming in plasma-enhanced catalytic reaction. Catalysis Today. 2015;**256**:67-75. DOI: 10.1016/j.cattod.2015.05.011

[11] Nozaki T, Okazaki K. Non-thermal plasma catalysis of methane: Principles, energy efficiency, and applications. Catalysis Today. 2013;**211**:29-38. DOI: 10.1016/j.cattod.2013.04.002

[12] Ray D, Reddy PMK, Subrahmanyam C. Ni-Mn/γ-Al$_2$O$_3$ assisted plasma dry reforming of methane. Catalysis Today. 2018;**309**:212-218. DOI: 10.1016/j.cattod.2017.07.003

[13] Zeng YX, Wang L, Wu CF, Wang JQ, Shen BX, Tu X. Low temperature reforming of biogas over K-, Mg- and Ce-promoted Ni/Al$_2$O$_3$ catalysts for the production of hydrogen rich syngas: Understanding the plasma-catalytic synergy. Applied Catalysis B: Environmental. 2018;**224**:469-478. DOI: 10.1016/j.apcatb.2017.10.017

[14] Tu X, Gallon HJ, Twigg MV, Gorry PA, Whitehead JC. Dry reforming of methane over a Ni/Al$_2$O$_3$ catalyst in a coaxial dielectric barrier discharge reactor. Journal of Physics D: Applied Physics. 2011;**44**:274007. DOI: 10.1088/0022-3727/44/27/274007

[15] Van Laer K, Bogaerts A. Influence of gap size and dielectric constant of the packing material on the plasma behaviour in a packed bed DBD reactor: A Fluid Modelling Study. Plasma Processes and Polymers.

2016;**14**:1600129. DOI: 10.1002/
ppap.201600129

[16] Michielsen I, Uytdenhouwen Y,
Pype J, Michielsen B, Mertens J,
Reniers F, et al. CO_2 dissociation in a
packed bed DBD reactor: First steps
towards a better understanding of
plasma catalysis. Chemical Engineering
Journal. 2017;**326**:477-488. DOI:
10.1016/j.cej.2017.05.177

[17] Mei D, Tu X. Atmospheric pressure
non-thermal plasma activation of
CO_2 in a packed-bed dielectric barrier
discharge reactor. Chemphyschem.
2017;**18**:3253-3259. DOI: 10.1002/
cphc.201700752

[18] Mei D, Tu X. Conversion of CO_2 in
a cylindrical dielectric barrier discharge
reactor: Effects of plasma processing
parameters and reactor design. Journal
of CO_2 Utilization. 2017;**19**:68-78. DOI:
10.1016/j.jcou.2017.02.015

[19] Uytdenhouwen Y, Van Alphen S,
Michielsen I, Meynen V, Cool P,
Bogaerts A. A packed-bed DBD micro
plasma reactor for CO_2 dissociation:
Does size matter? Chemical Engineering
Journal. 2018;**348**:557-568. DOI:
10.1016/j.cej.2018.04.210

[20] Mustafa MF, Fu X, Liu Y, Abbas Y,
Wang H, Lu W. Volatile organic
compounds (VOCs) removal in nonthermal plasma double dielectric barrier
discharge reactor. Journal of Hazardous
Materials. 2018;**347**:317-324. DOI:
10.1016/j.jhazmat.2018.01.021

[21] Veerapandian S, Leys C, De
Geyter N, Morent R. Abatement of
VOCs using packed bed non-thermal
plasma reactors: A review. Catalysts.
2017;**7**:113. DOI: 10.3390/catal7040113

[22] Xiao G, Xu W, Wu R, Ni M, Du C,
Gao X, et al. Non-thermal plasmas for
VOCs abatement. Plasma Chemistry
and Plasma Processing. 2014;**34**:1033-
1065. DOI: 10.1007/s11090-014-9562-0

[23] Yao S, Shen X, Zhang X, Han J,
Wu Z, Tang X, et al. Sustainable
removal of particulate matter
from diesel engine exhaust at low
temperature using a plasma-catalytic
method. Chemical Engineering Journal.
2017;**327**:343-350. DOI: 10.1016/j.
cej.2017.06.122

[24] Zhu X, Gao X, Qin R, Zeng Y, Qu R,
Zheng C, et al. Plasma-catalytic removal
of formaldehyde over Cu–Ce catalysts
in a dielectric barrier discharge reactor.
Applied Catalysis B: Environmental.
2015;**170**:293-300. DOI: 10.1016/j.
apcatb.2015.01.032

[25] Patil BS, Cherkasov N, Lang J,
Ibhadon AO, Hessel V, Wang Q. Low
temperature plasma-catalytic NO_x
synthesis in a packed DBD reactor:
Effect of support materials and
supported active metal oxides.
Applied Catalysis B: Environmental.
2016;**194**:123-133. DOI: 10.1016/j.
apcatb.2016.04.055

[26] Kameshima S, Mizukami R,
Yamazaki T, Prananto LA, Nozaki
T. Interfacial reactions between DBD
and porous catalyst in dry methane
reforming. Journal of Physics D:
Applied Physics. 2018;**51**:114006. DOI:
10.1088/1361-6463/aaad7d

[27] Kim HH, Teramoto Y, Negishi N,
Ogata A. A multidisciplinary approach
to understand the interactions of
nonthermal plasma and catalyst: A
review. Catalysis Today. 2015;**256**:13-22.
DOI: 10.1016/j.cattod.2015.04.009

[28] Du Y, Tamura K, Moore S,
Peng Z, Nozaki T, Bruggeman
PJ. $CO(B^1\Sigma^+ \rightarrow A^1\Pi)$ angstrom system for
gas temperature measurements in CO_2
containing plasmas. Plasma Chemistry
and Plasma Processing. 2017;**37**:29-41.
DOI: 10.1007/s11090-016-9759-5

[29] Bröer S, Hammer T. Selective
catalytic reduction of nitrogen oxides
by combining a non-thermal plasma
and a V_2O_5-WO_3/TiO_2 catalyst.

Applied Catalysis B: Environmental. 2000;**28**:101-111. DOI: 10.1016/ S0926-3373(00)00166-1

[30] Futamura S, Zhang A, Einaga H, Kabashima H. Involvement of catalyst materials in nonthermal plasma chemical processing of hazardous air pollutants. Catalysis Today. 2002;**72**:259-265. DOI: 10.1016/ S0920-5861(01)00503-X

[31] Roland U, Holzer F, Kopinke FD. Improved oxidation of air pollutants in a non-thermal plasma. Catalysis Today. 2002;**73**:315-323. DOI: 10.1016/S0920-5861(02)00015-9

[32] Chang CL, Lin TS. Elimination of carbon monoxide in the gas streams by dielectric barrier discharge systems with Mn catalyst. Plasma Chemistry and Plasma Processing. 2005;**25**:387-401. DOI: 10.1007/s11090-004-3135-6

[33] Van Durme J, Dewulf J, Sysmans W, Leys C, Van Langenhove H. Efficient toluene abatement in indoor air by a plasma catalytic hybrid system. Applied Catalysis B: Environmental. 2007;**74**:161-169. DOI: 10.1016/j. apcatb.2007.02.006

[34] Wallis AE, Whitehead JC, Zhang K. Plasma-assisted catalysis for the destruction of CFC-12 in atmospheric pressure gas streams using TiO_2. Catalysis Letters. 2007;**113**:29-33. DOI: 10.1007/s10562-006-9000-x

[35] Huu TP, Gil S, Da Costa P, Giroir-Fendler A, Khacef A. Plasma-catalytic hybrid reactor: Application to methane removal. Catalysis Today. 2015;**257**:86-92. DOI: 10.1016/j. cattod.2015.03.001

[36] Wang Q, Yan B-H, Jin Y, Cheng Y. Dry reforming of methane in a dielectric barrier discharge reactor with Ni/Al_2O_3 catalyst: Interaction of catalyst and plasma. Energy & Fuels. 2009;**23**:4196-4201. DOI: 10.1021/ ef900286j

[37] Kim HH, Tsubota S, Daté M, Ogata A, Futamura S. Catalyst regeneration and activity enhancement of Au/TiO_2 by atmospheric pressure nonthermal plasma. Applied Catalysis A: General. 2007;**329**:93-98. DOI: 10.1016/j.apcata.2007.06.029

[38] Kim HH, Sugasawa M, Hirata H, Teramoto Y, Kosuge K, Negishi N, et al. Ozone-assisted catalysis of toluene with layered ZSM-5 and Ag/ZSM-5 zeolites. Plasma Chemistry and Plasma Processing. 2013;**33**:1083-1098. DOI: 10.1007/s11090-013-9487-z

[39] Chavadej S, Kiatubolpaiboon W, Rangsunvigit P, Sreethawong T. A combined multistage corona discharge and catalytic system for gaseous benzene removal. Journal of Molecular Catalysis A: Chemical. 2007;**263**:128-136. DOI: 10.1016/j. molcata.2006.08.061

[40] Sreethawong T, Thakonpatthanakun P, Chavadej S. Partial oxidation of methane with air for synthesis gas production in a multistage gliding arc discharge system. International Journal of Hydrogen Energy. 2007;**32**:1067-1079. DOI: 10.1016/j.ijhydene.2006.07.013

[41] Harling AM, Glover DJ, Whitehead JC, Zhang K. Novel method for enhancing the destruction of environmental pollutants by the combination of multiple plasma discharges. Environmental Science & Technology. 2008;**42**:4546-4550. DOI: 10.1021/es703213p

[42] Dombrowski E, Peterson E, Del Sesto D, Utz AL. Precursor-mediated reactivity of vibrationally hot molecules: Methane activation on Ir (111). Catalysis Today. 2015;**244**:10-18. DOI: 10.1016/j. cattod.2014.10.025

[43] Nozaki T, Muto N, Kado S, Okazaki K. Dissociation of vibrationally excited methane on Ni catalyst: Part

1. Application to methane steam reforming. Catalysis Today. 2004;**89**: 57-65. DOI: 10.1016/j.cattod.2003.11.040

[44] Nozaki T, Tsukijihara H, Fukui W, Okazaki K. Kinetic analysis of the catalyst and nonthermal plasma hybrid reaction for methane steam reforming. Energy & Fuels. 2007;**21**(5):2525-2530. DOI: 10.1021/ef070117+

[45] Neyts EC. Plasma-surface interactions in plasma catalysis. Plasma Chemistry and Plasma Processing. 2016;**36**:185-212. DOI: 10.1007/s11090-015-9662-5

[46] Chen HL, Lee HM, Chen SH, Chao Y, Chang MB. Review of plasma catalysis on hydrocarbon reforming for hydrogen production—Interaction, integration, and prospects. Applied Catalysis B: Environmental. 2008;**85**:1-9. DOI: 10.1016/j.apcatb.2008.06.021

[47] Van Durme J, Dewulf J, Leys C, Van Langenhove H. Combining non-thermal plasma with heterogeneous catalysis in waste gas treatment: A review. Applied Catalysis B: Environmental. 2008;**78**:324-333. DOI: 10.1016/j.apcatb.2007.09.035

[48] Neyts EC, Bogaerts A. Understanding plasma catalysis through modelling and simulation—A review. Journal of Physics D: Applied Physics. 2014;**47**:224010. DOI: 10.1088/0022-3727/47/22/224010

[49] Vandenbroucke AM, Morent R, De Geyter N, Leys C. Non-thermal plasmas for non-catalytic and catalytic VOC abatement. Journal of Hazardous Materials. 2011;**195**:30-54. DOI: 10.1016/j.jhazmat.2011.08.060

[50] Chang JS, Kostov KG, Urashima K, Yamamoto T, Okayasu Y, Kato T, et al. Removal of NF_3 from semiconductor-process flue gases by tandem packed-bed plasma and adsorbent hybrid systems. IEEE Transactions on Industry Applications. 2000;**36**:1251-1259. DOI: 10.1109/28.871272

[51] Kang WS, Park JM, Kim Y, Hong SH. Numerical study on influences of barrier arrangements on dielectric barrier discharge characteristics. IEEE Transactions on Plasma Science. 2003;**31**:504-510. DOI: 10.1109/TPS.2003.815469

[52] Mizuno A, Ito H. Basic performance of an electrostatically augmented filter consisting of a packed ferroelectric pellet layer. Journal of Electrostatics. 1990;**25**:97-107. DOI: 10.1016/0304-3886(90)90039-X

[53] Mizuno A, Yamazaki Y, Obama S, Suzuki E, Okazaki K. Effect of voltage waveform on partial discharge in ferroelectric pellet layer for gas cleaning. IEEE Transactions on Industry Applications. 1993;**29**:262-267. DOI: 10.1109/28.216530

[54] Kim HH, Teramoto Y, Ogata A. Time-resolved imaging of positive pulsed corona-induced surface streamers on TiO_2 and γ-Al_2O_3-supported Ag catalysts. Journal of Physics D: Applied Physics. 2016;**49**(41):415204. DOI: 10.1088/0022-3727/49/45/459501

[55] Kim HH, Kim JH, Ogata A. Microscopic observation of discharge plasma on the surface of zeolites supported metal nanoparticles. Journal of Physics D: Applied Physics. 2009;**42**:135210. DOI: 10.1088/0022-3727/42/13/135210

[56] Zhang YR, Van Laer K, Neyts EC, Bogaerts A. Can plasma be formed in catalyst pores? A modeling investigation. Applied Catalysis B: Environmental. 2016;**185**:56-67. DOI: 10.1016/j.apcatb.2015.12.009

[57] Hensel K, Martišovitš V, Machala Z, Janda M, Leštinský M, Tardiveau P, et al. Electrical and optical properties of AC microdischarges in porous ceramics. Plasma Processes and Polymers. 2007;**4**:682-693. DOI: 10.1002/ppap.200700022

[58] Wang Z, Zhang Y, Neyts EC, Cao X, Zhang X, Jang BWL, et al. Catalyst preparation with plasmas: How does it work? ACS Catalysis. 2018;**8**:2093-2110. DOI: 10.1021/acscatal.7b03723

[59] Blin-Simiand N, Tardiveau P, Risacher A, Jorand F, Pasquiers S. Removal of 2-heptanone by dielectric barrier discharges—The effect of a catalyst support. Plasma Processes and Polymers. 2005;**2**:256-262. DOI: 10.1002/ppap.200400088

[60] Liu C, Zou J, Yu K, Cheng D, Han Y, Zhan J, et al. Plasma application for more environmentally friendly catalyst preparation. Pure and Applied Chemistry. 2006;**78**:1227-1238. DOI: 10.1351/pac200678061227

[61] Hong J, Chu W, Chernavskii PA, Khodakov AY. Cobalt species and cobalt-support interaction in glow discharge plasma-assisted Fischer-Tropsch catalysts. Journal of Catalysis. 2010;**273**:9-17. DOI: 10.1016/j.jcat.2010.04.015

[62] Shang S, Liu G, Chai X, Tao X, Li X, Bai M, et al. Research on Ni/γ-Al$_2$O$_3$ catalyst for CO$_2$ reforming of CH$_4$ prepared by atmospheric pressure glow discharge plasma jet. Catalysis Today. 2009;**148**:268-274. DOI: 10.1016/j.cattod.2009.09.011

[63] Wang L, Zhao Y, Liu C, Gong W, Guo H. Plasma driven ammonia decomposition on a Fe-catalyst: Eliminating surface nitrogen poisoning. Chemical Communications. 2013;**49**:3787-3789. DOI: 10.1039/C3CC41301B

[64] Smith RR, Killelea DR, DelSesto DF, Utz AL. Preference for vibrational over translational energy in a gas-surface reaction. Science. 2004;**304**:992-995. DOI: 10.1126/science.1096309

[65] Demidyuk V, Whitehead JC. Influence of temperature on gas-phase toluene decomposition in plasma-catalytic system. Plasma Chemistry and Plasma Processing. 2007;**27**:85-94. DOI: 10.1007/s11090-006-9045-z

[66] Nozaki T, Okazaki K. Carbon nanotube synthesis in atmospheric pressure glow discharge: A review. Plasma Processes and Polymers. 2008;**5**:300-321. DOI: 10.1002/ppap.200700141

[67] Nozaki T, Fukui W, Okazaki K. Reaction enhancement mechanism of the nonthermal discharge and catalyst hybrid reaction for methane reforming. Energy & Fuels. 2008;**22**:3600-3604. DOI: 10.1021/ef800461k

[68] Mutz B, Carvalho HWP, Mangold S, Kleist W, Grunwaldt JD. Methanation of CO$_2$: Structural response of a Ni-based catalyst under fluctuating reaction conditions unraveled by operando spectroscopy. Journal of Catalysis. 2015;**327**:48-53. DOI: 10.1016/j.jcat.2015.04.006

[69] Mutz B, Carvalho HWP, Kleist W, Grunwaldt JD. Dynamic transformation of small Ni particles during methanation of CO$_2$ under fluctuating reaction conditions monitored by operando X-ray absorption spectroscopy. Journal of Physics: Conference Series. 2016;**712**:012050. DOI: 10.1088/1742-6596/712/1/012050

[70] Snoeckx R, Bogaerts A. Plasma technology—A novel solution for CO$_2$ conversion? Chemical Society Reviews. 2017;**46**:5805-5863. DOI: 10.1039/C6CS00066E

[71] Belov I, Vanneste J, Aghaee M, Paulussen S, Bogaerts A. Synthesis of micro- and nanomaterials in CO$_2$ and CO dielectric barrier discharges. Plasma Processes and Polymers. 2016;**14**:1600065. DOI: 10.1002/ppap.201600065

[72] Aerts R, Somers W, Bogaerts A. Carbon dioxide splitting in a dielectric

barrier discharge plasma: A combined experimental and computational study. ChemSusChem. 2015;**8**:702-716. DOI: 10.1002/cssc.201402818

[73] Butterworth T, Elder R, Allen R. Effects of particle size on CO_2 reduction and discharge characteristics in a packed bed plasma reactor. Chemical Engineering Journal. 2016;**293**:55-67. DOI: 10.1016/j.cej.2016.02.047

[74] Mutz B, Gänzler MA, Nachtegaal M, Müller O, Frahm R, Kleist W, et al. Surface oxidation of supported Ni particles and its impact on the catalytic performance during dynamically operated methanation of CO_2. Catalysts. 2017;**7**:279. DOI: 10.3390/catal7090279

[75] Foppa L, Margossian T, Kim SM, Müller C, Copéret C, Larmier K, et al. Contrasting the role of Ni/Al_2O_3 interfaces in water–gas shift and dry reforming of methane. Journal of the American Chemical Society. 2017;**139**:17128-17139. DOI: 10.1021/jacs.7b08984

[76] Silaghi MC, Comas-Vives A, Copéret C. CO_2 activation on Ni/γ-Al_2O_3 catalysts by first-principles calculations: From ideal surfaces to supported nanoparticles. ACS Catalysis. 2016;**6**:4501-4505. DOI: 10.1021/acscatal.6b00822

[77] Bogaerts A, Kozák T, van Laer K, Snoeckx R. Plasma-based conversion of CO_2: Current status and future challenges. Faraday Discussions. 2015;**183**:217-232. DOI: 10.1039/C5FD00053J

[78] Van Rooij GJ, van den Bekerom DCM, den Harder N, Minea T, Berden G, Bongers WA, et al. Taming microwave plasma to beat thermodynamics in CO_2 dissociation. Faraday Discussions. 2015;**183**:233-248. DOI: 10.1039/C5FD00045A

[79] Nozaki T, Miyazaki Y, Unno Y, Okazaki K. Energy distribution and heat transfer mechanisms in atmospheric pressure non-equilibrium plasmas. Journal of Physics D: Applied Physics. 2001;**34**:3383

[80] Nozaki T, Unno Y, Okazaki K. Thermal structure of atmospheric pressure non-equilibrium plasmas. Plasma Sources Science and Technology. 2002;**11**:431. DOI: 10.1088/0963-0252/11/4/310

Chapter 4

Progress in Plasma-Assisted Catalysis for Carbon Dioxide Reduction

Guoxing Chen, Ling Wang,
Thomas Godfroid and Rony Snyders

Abstract

Production of chemicals and fuels based on CO_2 conversion is attracting a special attention nowadays, especially regarding the fast depletion of fossil resources and increase of CO_2 emissions into the Earth's atmosphere. Recently, plasma technology has gained increasing interest as a non-equilibrium medium suitable for CO_2 conversion, which provides a promising alternative to the conventional pathway for greenhouse gas conversion. The combination of plasma and catalysis is of great interest for turning plasma chemistry in applications related to pollution and energy issues. In this chapter a short review of the current progress in plasma-assisted catalytic processes for CO_2 reduction is given. The most widely used discharges for CO_2 conversion are presented and briefly discussed, illustrating how to achieve a better energy and conversion efficiency. The chapter includes the recent status and advances of the most promising candidates (plasma catalysis) to obtain efficient CO_2 conversion, along with the future outlook of this plasma-assisted catalytic process for further improvement.

Keywords: green energy, plasma-based CO_2 conversion, plasma catalysis, oxygen vacancies, synergistic effect

1. Introduction

The utilization of CO_2 for production of fuels, energy storage media, chemicals or aggregates is attracting interest worldwide due to the essential contribution of the greenhouse gases to the global warming. CO_2 capture and utilization are considered as a promising option for the mitigation of CO_2 emissions, which provides a lower carbon footprint for the synthesis of value-added products than those produced by conventional processes using fossil fuels. In spite of the continuously increasing interest for CO_2 recycling, there are significant challenges to overcome due to its stable molecular structure and low chemical activity. There are several methods that can be used to convert CO_2, including traditional catalysis, photochemical, biochemical, solar thermochemical, electrochemical and plasma chemical. Snoeckx and Bogaerts recently made a detailed comparison of these technologies as shown in **Table 1** [1]. They concluded that the plasma technology fares very well in this comparison and is quite promising. Indeed, nonthermal plasma has attracted much attention of the scientific community as a non-equilibrium medium suitable for CO_2

	Use of rare earth metals	Renewable energy	Turnkey process	Conversion and yield	Separation step needed	Oxygenated products (e.g. alcohols, acids)	Investment cost	Operating cost	Overall flexibility
Traditional catalysis	Yes	-	No	High	Yes	Yes	Low	High	Low
Catalysis by MW-heating	Yes	Indirect	No	High	Yes	Yes	Low	Low	Low
Electro-chemical	Yes	Indirect	No[b]	High	Yes[c]	Yes	Low	Low	Medium
Solar thermo-chemical	Yes	Direct	NA	High	No	No	High	Low	Low
Photo-chemical	Yes	Direct[a]	Yes	Low	Yes	Yes	Low	Low	Low
Biochemical	No	Direct[a]	No	Medium	Yes[d]	Yes	High /low	High	Low
Plasma-chemical	No	Indirect	Yes	High	Yes[e]	Yes	Low	Low	High

[a]*Bio- and photochemical processes can also rely on indirect renewable energy when they are coupled with artificial lighting.*
[b]*Electrochemical cells are turnkey, but generally the cells need to operate at elevated temperatures and the cells are sensitive to on/off fluctuations.*
[c]*The need for post-reaction separation for the electrochemical conversion highly depends on the process and cell type used.*
[d]*Biochemical CO_2 conversion requires very energy-intensive post-reaction separation and processing steps.*
[e]*The need for post-reaction separation for plasma technology highly depends on the process.*

Table 1.
Comparing the advantages and disadvantages of the different technologies for CO_2 reduction (adapted from [1]).

conversion, which provides an attractive alternative to the conventional pathway for CO_2 recycling, such as traditional catalysis and solar thermochemical process.

Nonthermal plasmas have been successfully utilized in many applications for the environmental control (such as gaseous pollutant abatement), material science (such as surface treatment) and medical applications (such as wound and cancer treatment) [1–3]. Nowadays, an increasing interest has been focused on examining their use for CO_2 utilization [3–54]. In comparison to the other processes, plasma process is fast: plasma has the potential to enable thermodynamically unfavorable chemical reactions (e.g. CO_2 dissociation) to occur on the basis of its non-equilibrium properties, low-power requirement and its capacity to induce physical and chemical reactions at a relatively low temperature. In addition, plasma can be ignited and shut off quickly, which enables plasma technology powered by renewable energy to act as an efficient chemical switch for the conversion purposes. Although plasma technology shows great potential, there is always a trade-off between the energy efficiency and conversion efficiency in plasma-only process. Last but not least, the conversion efficiency can be significantly improved by combining plasma with catalyst while maintaining high-energy efficiency.

Plasma catalysis (also referred to as plasma-enhanced catalysis, plasma-driven catalysis or plasma-assisted catalysis) has gathered attention as a way of increasing energy efficiency and optimizing the byproduct distribution [55]. On one hand, the catalyst can increase reaction rates and overall process selectivity. The nonthermal plasma can provide energy to drive highly endothermic processes. Plasma-catalytic processes have great potential to reduce the activation barrier of different reactions and improve the conversion rates. In addition, the nonthermal plasma itself can influence the acid–base nature of the supports, enhance the dispersion of the

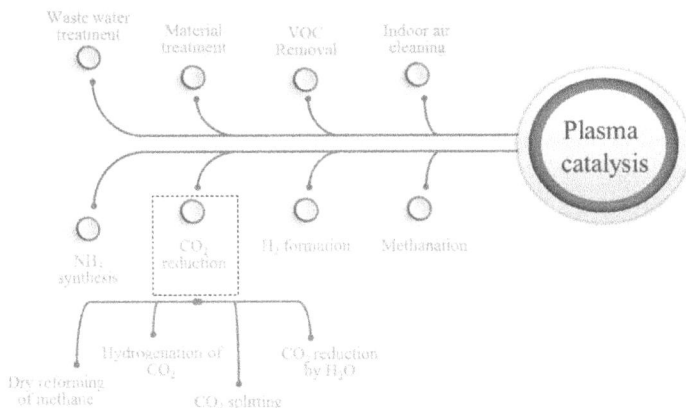

Figure 1.
Applications of plasma catalysis.

supported metals and even adjust the microstructure of the metal nanoparticles and metal-support interface [56, 57] and in this way change the catalyst properties. All these factors contribute in different ways to the enhancement of energy efficiency of the plasma process as well as the catalyst stability, due to a synergy that occurs between the catalyst and the plasma [58]. This novel technique combines the advantages of high product selectivity from thermal catalysis and the fast startup from plasma technique. Plasma catalysis has been widely investigated for many applications. **Figure 1** briefly summarizes the main application areas of plasma catalysis. In the domain of energy applications, the use of plasma catalysis for dry reforming, CO_2 reduction, hydrogen production, methanation and ammonia (NH_3) synthesis has been intensively studied. In this chapter, however, we focus only on their application for CO_2 conversion into value-added chemicals and fuels.

2. Brief theoretical background

2.1 CO_2 dissociation chemistry

As mentioned in Introduction, nonthermal plasma shows a great potential for an efficient CO_2 utilization. Different routes for CO_2 conversion have been investigated using plasma-catalytic process. **Table 2** summarizes some of the main reactions

Process	Reaction	Enthalpy (ΔH) kJ mol^{-1}	Enthalpy (ΔH) eV/molecule
CO_2 splitting	$CO_2 \rightarrow CO + \frac{1}{2}O_2$	279.8	2.9
Dry reforming of methane	$CO_2 + CH_4 \rightarrow 2CO + 2H_2$	247.4	2.6
Methanol synthesis	$CO_2 + 3H_2 \rightarrow CH_3OH + H_2O$	−128	−1.3
Methanation	$CO_2 + 4H_2 \rightarrow CH_4 + 2H_2O$	−164.8	−1.7
Reverse water-gas shift reaction	$CO_2 + H_2 \rightarrow CO + H_2O$	41.2	0.4
Water-gas shift reaction	$CO + H_2O \rightarrow CO_2 + H_2$	−41.2	−0.4
Methanation	$CO + 3H_2 \rightarrow CH_4 + H_2O$	−205.8	−2.1
Water spitting	$H_2O \rightarrow H_2 + \frac{1}{2}O_2$	250.9	2.6

Table 2.
Chemical reactions related to CO₂ reduction and their enthalpies.

usually considered in plasma chemistry for CO_2 reduction using different pathways (such as dry reforming of methane, hydrogenation of CO_2). Significant attention has been given to plasma-catalytic dry reforming of methane (DRM) using supported Ni catalysts. However, most of these studies focused primarily on identifying plasma-catalytic chemical reactions to maximize process performance. Optical emission spectroscopy and plasma chemical kinetic modeling should be used to achieve a better understanding on the formation of a wide range of reactive species in this plasma-catalytic reforming process. Recently, Chung et al. had described the mechanisms of catalysis promotion, elucidated the synergistic effects between catalyst and plasma and proposed possible approaches to optimize DRM process performance [2]. As explained by Fridman [3], cumulative vibrational excitations of the CO_2 molecule can result in a highly energy-efficient stepwise dissociation. Thus, CO_2 splitting using nonthermal plasmas has been considered as another promising pathway to produce synthetic fuels via CO, as an intermediate product. As well-accepted in the literature, dissociation of a CO_2 molecule in plasma is represented by the following global reaction [3]:

$$CO_2 \rightarrow CO + \tfrac{1}{2}O_2, \Delta H = 2.9 \text{ eV/molecule} \tag{1}$$

The main pathways for decomposition of CO_2 molecule include the electron impact dissociation:

$$CO_2 \rightarrow CO + O, \Delta H = 5.5 \text{ eV/molecule} \tag{2}$$

which is often accompanied by the further recombination of atomic O:

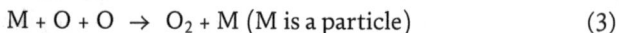

$$M + O + O \rightarrow O_2 + M \text{ (M is a particle)} \tag{3}$$

In addition to this, the vibrationally excited CO_2 molecules may also undergo decomposition via the collisions with atomic O:

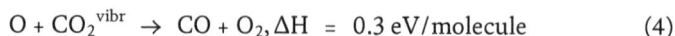

$$O + CO_2{}^{vibr} \rightarrow CO + O_2, \Delta H = 0.3 \text{ eV/molecule} \tag{4}$$

as well as with the plasma electrons:

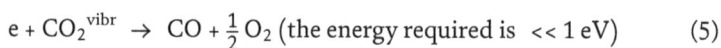

$$e + CO_2{}^{vibr} \rightarrow CO + \tfrac{1}{2}O_2 \text{ (the energy required is } << 1 \text{ eV)} \tag{5}$$

Traditionally, to characterize the process efficiency, two main parameters reflecting the *conversion* efficiency and *energy* efficiency are used. The conversion efficiency (χ) and energy efficiency (η) of CO_2 are defined as follows:

$$\chi = \frac{\text{moles of } CO_2 \text{ input} - \text{moles of } CO_2 \text{ output}}{\text{moles of } CO_2 \text{ input}} \tag{6}$$

$$\eta = \frac{\chi * 2.9 \text{ eV}}{SEI} \tag{7}$$

Here the specific energy input (SEI) per molecule is given by the ratio of the discharge power (P) to the gas flow rate (F) through the discharge volume.

2.2 Plasma catalysis

When catalysts are combined with plasmas, they can be classified into three systems, i.e. single stage, two stage, and multistage, depending on the location of

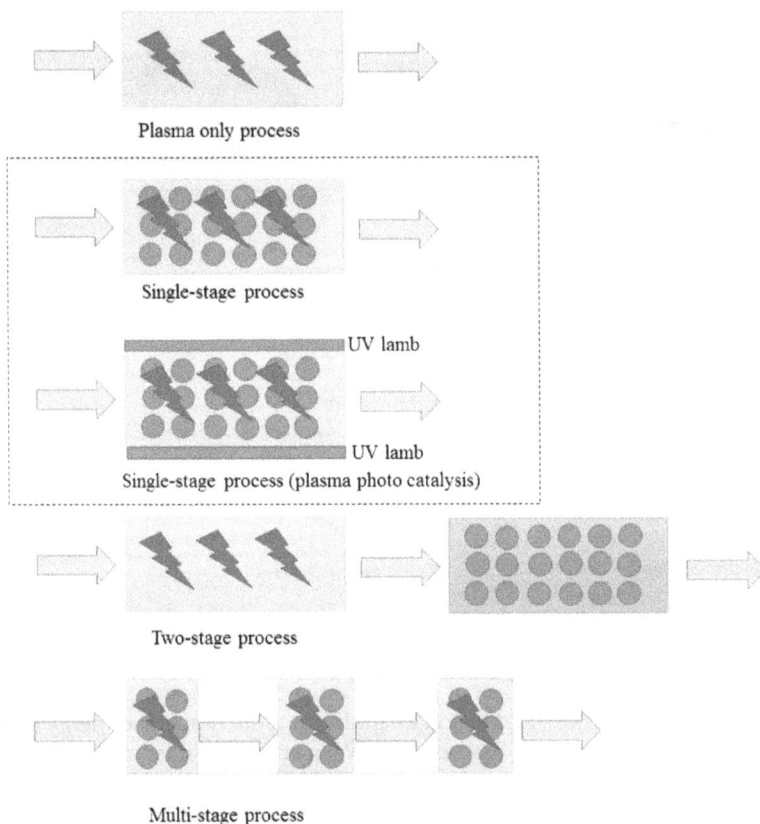

Figure 2.
Schematic diagram of different plasma-catalyst configurations according to the catalyst bed position and number.

the catalyst [59, 60]. These three configurations are illustrated in **Figure 2**. In all cases, the plasma can be used to supply energy for catalyst activation, and it can also provide the reactive gas species needed for reactions on the catalyst surface. The single-stage type is constructed by coating catalyst on the surface of electrode(s) or packing catalyst within the plasma zone, which is also called in-plasma catalysis (IPC). The catalysts could completely or just partially overlap with the plasma zone. In this manner, the plasma and catalysis could directly interact with each other. This single-stage system is also easy to combine with the UV irradiation, which is known as plasma photo catalysis, as shown in **Figure 2**. For the two-stage type, the catalyst is placed after the plasma discharge region; it is also called post-plasma catalysis (PPC). The plasma provides chemically reactive species for catalysis or pre-converts reactants into the easier-to-convert products to accelerate the catalysis. In the nonthermal plasma catalysis system, the long-lived reactive species produced by plasma, e.g. vibration-excited species, radicals, and ionized molecules, can react with the catalyst to induce catalytic reactions via either the Eley-Rideal mechanism or Langmuir-Hinshelwood mechanism [2, 59]. The multistage plasma catalysis system is a promising option for the industrial use in the future. Different functions of the catalysts can be combined to achieve certain expected reaction in the multistage system.

In the context of plasma catalysis, the synergy is referring to a surplus effect of combining the plasma with catalyst, namely, when the resulting effect has a higher

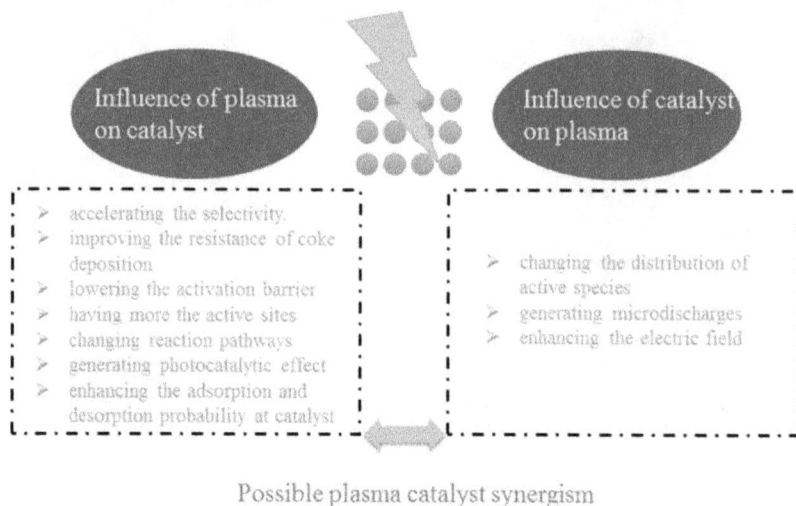

Possible plasma catalyst synergism

Figure 3.
Interaction between catalyst and plasma.

impact than the sum of their individual impacts. In several studies, the combination of plasma and catalysts has been found to have synergistic effects [34, 35, 61, 62]. A highly important synergistic effect of plasma catalysis is promotion of catalyst activity at reduced temperatures, and hence, a significant reduction in the energy cost for activating the catalyst [34]. For example, Wang et al. illustrated such synergy for plasma catalysis of dry reforming methane (DRM) in the single-stage system with Ni/Al_2O_3 catalyst but did not observe this synergy in the two-stage system or when the catalyst is only placed at the end of the plasma zone [62]. Typical synergistic effect factors of 1.25–1.5 were obtained. Zhang et al. presented the results on the plasma-catalyst synergy in the case of dry reforming methane using different $Cu-Ni/\gamma-Al_2O_3$ catalysts [63]. The effect was observed on the conversions of CH_4 and CO_2, where the result for the plasma-catalytic reaction was greater than the sum of the catalyst-only or plasma-only results. The selectivity towards H_2 and CO production was also enhanced by the use of plasma catalysis. In general, the enhanced performance of plasma catalysis can in part be attributed to vibrational excitation of CO_2 in the plasma, which enables easier dissociation at low temperature on the catalyst surface. The plasma electrons in turn affect the catalyst properties (chemical composition or catalytic structure). Synergistic effects in the plasma and catalyst are illustrated in **Figure 3**. Plasma can alter the physicochemical characteristics of catalyst via several routes, which are induced mainly by energetic electron generation. In the meantime, a catalyst can induce electric field concentration due to its pore structure and dielectric properties. Hence, both electric field distribution and catalyst characteristics are modified to have better DRM performance.

3. Plasma-assisted catalytic conversion of CO_2

Nonthermal plasma technology provides an attractive alternative to the other (classical) technologies for converting inert carbon emissions. Different types of plasmas have already been used for CO_2 reduction, including dielectric barrier discharges (DBDs), glow discharges, radio frequency (RF) discharges, microwave (MW) discharges and gliding arc plasma (GAP) and corona discharges [8–39]. In this section, the most widely used discharges for CO_2 conversion are presented.

DBDs have been known for more than a century. They were first reported in 1857 by Siemens for the use in ozone production and were originally called 'silent' discharges [64]. The DBD is the most widely used discharge type for CO_2 conversion among the variety of other plasma sources because it is easy to handle with relatively cheap equipment and it operates at atmospheric pressure [34]. Even though the conversion efficiencies obtained in DBDs are generally quite low [1, 4, 5, 8, 9, 12, 13], the possibility to work at atmospheric pressure under non-equilibrium conditions is still a very strong advantage of these discharges. Combined with plasma catalysis, these discharges should also improve the selective production of the targeted compounds.

An atmospheric pressure GAP discharge can be formed between two flat knife-shaped electrodes with a gas flowing between them. These discharges are suitable for applications that require relatively large gas flows (several l/min). The gliding arc plasma can be operated in the thermal and nonthermal regime depending on the applied power and flow rate. Furthermore, the arc can be operated in the transition regime, which is an evolving arc starting in the thermal regime going to the nonthermal regime. This transition regime makes the discharge energy efficient for gas treatment. An energy efficiency of 43% was reported by Nunnally et al. for the decomposition of CO_2 in a reverse vortex flow gliding arc discharge, which is quite high compared to the efficiency obtained with DBDs (about 10%) [31]. The high level of efficiency can be attributed to non-equilibrium vibrational excitation of CO_2 and a high-temperature gradient between the gliding arc and the surrounding gas that results in fast quenching.

Plasmas generated by the injection of microwave power, i.e. electromagnetic radiation in the frequency range of 100 MHz–10 GHz, are called MW plasmas [65]. MW discharges are commonly generated using frequencies of 2.45 and 0.915 GHz. They can be operated over a wide pressure range (from few mTorr to the atmospheric pressure). The properties of the MW discharges operating at atmospheric pressure are close to those of thermal plasma. However, the MW discharges are far from thermodynamic equilibrium at low pressure. The performance of a microwave discharge in terms of efficiency of CO_2 dissociation process depends heavily on the plasma parameters such as power and operating pressure. The highest energy efficiency (about 90%) for pure CO_2 conversion was reported in a MW plasma operating with supersonic gas flows [22]. The ability to create a strong non-equilibrium environment in microwave discharges possesses highly vibrational states of CO_2 molecules, which are energy-efficient for CO_2 decomposition [3]. In general, the high efficiency of microwave plasmas is attained due to the high absorption of the applied power by electrons as well as relatively high excitation of the CO_2 asymmetric mode [24], which plays a key role for CO_2 decomposition [22]. In the low-pressure case, the microwave plasmas are typically characterized by an electron temperature around 1–2 eV and a gas temperature below 1500 K. Under these conditions, it has been estimated that about 95% of all the discharge energy is transferred from the plasma electrons to the CO_2 molecules, mostly to their asymmetric vibrational mode [3, 24].

Bogaerts et al. has presented some insights into how the electron energy is transferred to different channels of excitation, ionization or dissociation of the CO_2 molecules [1, 66]. **Figure 4** illustrates the fractional energy transferred from electrons to different channels of excitation, ionization and dissociation of CO_2, as a function of the reduced electric field (E/n) in a discharge. This plot is calculated based on the cross sections of the corresponding electron impact reactions [1, 37, 66]. In microwave plasma, the reduced electric field is typically around 50 Td, which is most appropriate for the vibrational excitation of CO_2. Fridman has shown that up to 97% of the total nonthermal discharge energy can be transferred from the plasma electrons to vibrational excitation of CO_2 molecules at an electron temperature around 1–2 eV or a reduced electric field (E/n) of about 20–40 Td [3, 36]. This is indeed indicated by the calculated curve referred to as the 'sum of all vibrations'

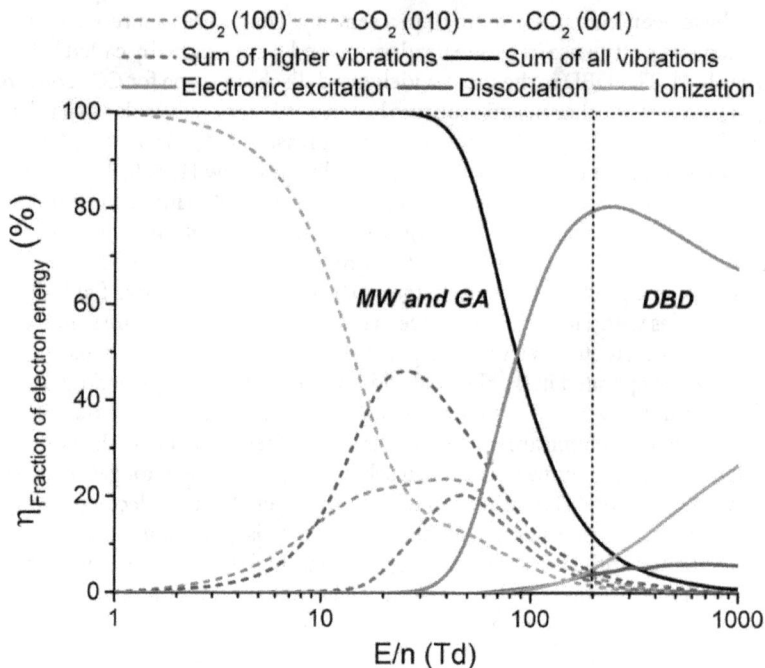

Figure 4.
The fraction of electron energy transferred to different channels of excitation as a function of the reduced electric field (E/n) (adapted from [1]).

shown in **Figure 4**. Moreover, the purple curve in **Figure 4** has its particular importance as it represents the first vibrational level of the asymmetric vibrational mode of CO_2, which represents the most important channel for the dissociation [66]. The energy efficiency for the dissociation of CO_2 is quite limited in a DBD plasma [3, 4, 11–13]. The electron temperature in a DBD is about 2–3 eV, which is somewhat high for efficient population of the CO_2 vibrational levels. The reduced electric field values are being typically about 200 Td or even higher, indicated as 'DBD region' in the figure. As a result of previous studies on CO_2 decomposition in plasma, it was concluded that higher pressures and lower values of reduced electric field make the vibrational excitation mechanism more favorable than the electronic excitation mechanism, explaining the higher energy efficiency of these types of discharges (e.g. MW, GAP) [1, 3, 22, 26, 28, 32, 33, 35, 37, 38].

3.1 MW region

In this chapter we have summarized the results from the recent publications on plasma set-ups with and without combining a catalyst for CO_2 conversion in **Table 2** and discussed the current research status on this topic. Porous Al_2O_3 (α-Al_2O_3 and γ-Al_2O_3) has been investigated in a pulsed corona discharge reactor for CO_2 conversion by Wen et al. [39]. γ-Al_2O_3 was found to enhance CO_2 conversion due to its high surface area and strong adsorption capability. Zhang et al. investigated CO_2 decomposition to CO and O_2 in a DBD reactor packed with a mixture of Ni/SiO_2 catalyst and $BaTiO_3$ spheres. In comparison to the reaction in the absence of a Ni/SiO_2 catalyst, introducing a Ni/SiO_2 catalyst to the plasma reactor packed with $BaTiO_3$ spheres slightly increase the CO_2 conversion from 19 to 23.5% at low temperatures [17]. Van Laer demonstrated a packing of ZrO_2 beads in a DBD reactor. The best combination of conversion (37.8%) and energy

(6.4%) efficiency was reached at a flow rate of 20 mL min^{-1} and an input power of 60 W [16]. Their simulation results suggest that the increased CO_2 conversion is caused by the presence of strong electric fields and thus high electron energies at the contact points, which thereby lowers the breakdown voltage. These findings suggest that the interactions between plasma and packing materials play an important role in the plasma conversion of CO_2. Brock et al. studied the catalytic effect of metallic coating on the decomposition of CO_2 in fan-type AC glow discharge plasma reactors, using a gas mixture of 2.5% CO_2 in He [19]. They showed that an Rh-coated reactor has the highest activity for the CO_2 decomposition compared to the reactors coated with Cu, Au, Pt and Pd and mixed rotor/stator systems (Rh/Au and Au/Rh).

In relation to microwave plasmas, Chen et al. reported that placing a NiO/TiO_2 catalyst in the downstream of a low-pressure microwave plasma significantly increased the CO_2 conversion efficiency and energy efficiency [25, 28]. They concluded that the oxygen vacancies provide the sites for adsorption of oxygen atoms from CO_2. The energetic electrons supplied by the plasma enhance the dissociative electron attachment of CO_2 at the surface. Recently, Ray et al. found that CO_2 conversion was enhanced upon packing CeO_2 into the discharge region of a DBD reactor. They also suggest this enhancement can be mainly attributed to the formation of oxygen vacancy defects on the surface of CeO_2, to stabilize the produced atomic oxygen, thereby preventing the revise reaction [18]. Spencer et al. experimentally investigated the conversion of CO_2 in an atmospheric pressure microwave plasma-catalytic system [21]. The results showed that Rh/TiO_2 coating on a monolithic cordierite structure used as a catalyst actually caused a drop in conversion efficiency due to reverse reactions occurring on the surface. Mei et al. demonstrated that the combination of plasma with $BaTiO_3$ and TiO_2 catalysts has a synergistic effect, which significantly enhances the conversion of CO_2 and the energy efficiency by a factor of 2.5 compared to the plasma reaction in the absence of a catalyst [7]. The overall synergistic effect resulting from the integration of DBD with catalysis for CO_2 conversion can be attributed to the dominant catalytic surface reaction driven by energetic electrons from the CO_2 discharge. Theoretical and experimental studies consistently showed that the CO_2 adsorption, activation and dissociation processes were significantly enhanced by the presence of oxygen vacancies [7, 23, 28, 67, 68]. The mechanism of plasma-catalytic CO_2 conversion can be described by **Figure 5**. The oxygen vacancies provide sites for the adsorption of

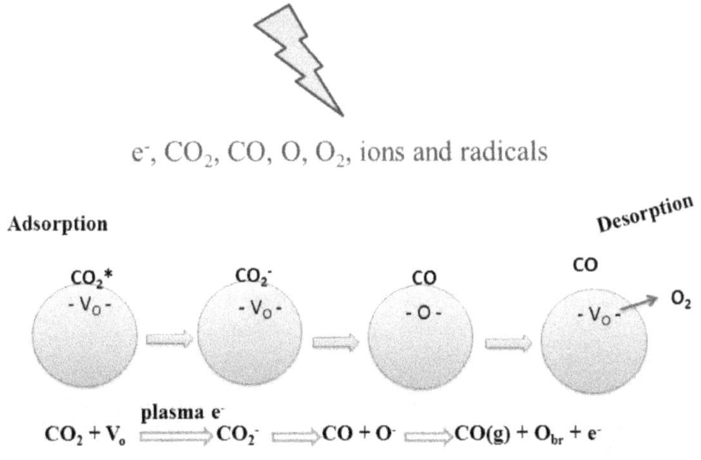

Figure 5.
Schematic mechanism of plasma-assisted catalytic process for CO_2 conversion.

Plasma type	Comments	Gas mixture	Catalyst	χ (%)	η (%)	SEI eV/ molecule	Ref.
DBD		CO_2	—	17	9	5.8	[4]
DBD		CO_2	—	30	1	87	[5]
DBD		CO_2	γ-Al_2O_3	20	4.9	12	[6]
DBD		CO_2	$BaTiO_3$	38	17	6.5	[7]
DBD		CO_2	—	18	4	13	[8]
DBD	Low flow rate	CO_2	—	14	8	5.2	[9]
DBD	10% CO_2 in the gas mixture	CO_2-H_2O-Ar	Ni/γ-Al_2O_3	36	23	4.5	[10]
DBD		CO2	—	35	2	50.8	[11]
DBD		CO_2	—	28.2	11.1	7.4	[12]
DBD		CO_2-N_2	—	4.5	4.5	2.9	[13]
DBD		CO_2	CaO	39.2	7.1	16	[14]
DBD		CO_2	—	20	10.4	5.6	[15]
DBD		CO_2	ZrO_2	2.9	9.6	9.6	[16]
DBD		CO_2	Ni/ SiO_2 + $BaTiO_3$	23.5	2.31	29.5	[17]
DBD		CO_2	CeO_2 (2 mm)	10.6	27.6	1.11	[18]
DBD		CO_2	TiO_2 (3–4 mm)	8.2	15.54	1.53	[18]
Glow		CO_2-Ar	Rh-coated	30	1.4	62	[19]
RF		CO_2	—	20	3	19	[20]
MW		CO_2-Ar	—	10	20	1.4	[21]
MW	Supersonic flow	CO_2	—	10	90	0.3	[22]
MW		CO_2	NiO/TiO_2	42	18	7.0	[23]
MW		CO_2-N_2	—	80	6	39	[24]
MW		CO_2	—	20	20	2.9	[25]
MW		CO_2	—	12	45	0.8	[26]
MW	CO_2:H_2O = 1:1	CO_2-H_2O	—	12	8.7	4	[27]
MW		CO_2	NiO/TiO_2	45	56	2.3	[28]
MW		CO_2	—				[29]
Corona		CO_2	—	11	2	16	[30]
Gliding arc		CO_2	—	4.6	43	0.3	[31]
Gliding arc		CO_2	—	15	19	2.3	[32]
Gliding arc		CO_2	—	10	34	0.85	[33]

Table 3.
Summary of the plasma-assisted catalytic CO_2 conversion for different discharge types.

oxygen atoms from CO_2. The energetic electrons supplied by the plasma enhance the dissociative electron attachment of CO_2 at the surface. Subsequently, CO desorbs or moves from the reactive site while the other O (bridging) atom 'heals' the oxygen vacancy. The oxygen vacancy can be regenerated via the recombination on the surface of a bridging oxygen atom with a gaseous oxygen atom. Such regeneration maintains the equilibrium of the active sites in the catalyst and controls the CO_2 conversion [23]. If the catalyst is placed in the plasma zone (single stage), the electron–hole pairs can be created by highly energetic electrons from the discharge upon the surface of photocatalysts once plasma can generate electrons of very similar energy (3–4 eV) to the photons. In this case, oxygen vacancy can be regenerated by oxidizing the surface O_2^- anions using holes, followed by releasing O_2 [7]. Plasma-catalytic conversion of CO_2 is a complex and challenging process involving a large number of physical and chemical reactions. The performance of the process is controlled by means of plasma parameters and the properties of the catalysts as well. This suggests that more systematic studies on both the plasma effects and the chemical effects of the catalyst are highly needed (**Table 3**).

4. Conclusions and perspectives

Plasma-assisted catalytic processes used for CO_2 reduction are gaining increasing interest worldwide. There is still a room, however, for further improvement of the CO_2 conversion and energy efficiencies through the optimization of the plasma parameters (e.g. high pressure and high flow rate) as well as through modification of catalysts.

The plasma-catalytic activities can be controlled by numerous factors such as the nature of the catalyst support, active metal sites, surface area and the nanoparticle size. Let us note that the catalyst preparation (sometime called 'activation') plays a very important role in this regard. In addition to these factors and also due to their existence, the fine-tuning of a given catalyst is inevitable and crucial factor for enhancing plasma-catalytic process efficiency. Several methods, such as loading different metal nanoparticles, using different catalyst preparation schemes (sol gel, co-precipitation, deposition-precipitation or hydrothermal synthesis), using larger surface area of the support, etc., can be mentioned to realize the mentioned tuning.

An important factor which cannot be omitted here is that a chosen catalyst material should have rather low costs to be potentially commercialized and implemented in the industrial scale. Moreover, as a result of recent development of the microwave discharges, namely, a possibility to place catalyst packing directly in the discharge zone can be a powerful way to take advantage of the stepwise vibrational excitation on the catalyst surface. In addition, using plasma as a tool for the preparation (activation) of the catalyst surface may be another promising way. To improve its application, a better insight into the underlying mechanisms of the plasma catalysis is desirable. A greater understanding of the plasma chemistry, both by plasma modeling and by coupling with other techniques such as catalysis and membrane materials, will allow this field to expand. We expect that the results presented in this chapter will provide useful insights into the plasma-assisted CO_2 conversion in the presence or the absence of catalysts, which may be used for greenhouse gas conversion in the industry.

Acknowledgements

The authors acknowledge financial support from the network on the Physical Chemistry of Plasma-Surface Interactions—Interuniversity Attraction Poles phase VII project (http://psi-iap7.ulb.ac.be/), supported by the Belgian Federal Office for Science Policy (BELSPO). The support of the 'REFORGAS GreenWin' project, grant No. 7267 (for GC, TG), should be acknowledged.

Author details

Guoxing Chen[1,2*], Ling Wang[4], Thomas Godfroid[3] and Rony Snyders[1,3]

1 Chemistry of Plasma Surface Interactions, University of Mons, Belgium

2 4MAT, Universite Libre de Bruxelles, Belgium

3 "Materia Nova" Research Center, Belgium

4 Institute for Materials Science, University of Stuttgart, Stuttgart, Germany

*Address all correspondence to: guoxchen@ulb.ac.be

IntechOpen

References

[1] Snoeckx R, Bogaerts A. Plasma technology – a novel solution for CO_2 conversion. Chemical Society Reviews. 2017;**46**:5805-5863

[2] Fridman A. Plasma Chemistry. London: Cambridge University Press; 2008

[3] Chung WC, Chang MB. Review of catalysis and plasma performance on dry reforming of CH_4 and possible synergistic effects. Renewable and Sustainable Energy Reviews. 2016;**62**:13-31

[4] Ozkan A, Dufour T, Bogaerts A, Reniers F. How do the barrier thickness and dielectric material influence the filamentary mode and CO_2 conversion in a flowing DBD? Plasma Sources Science and Technology. 2016;**25**(4):045016

[5] Paulussen S, Verheyde B, Xin T, Bie CD, Martens T, Petrovic D, et al. Conversion of carbon dioxide to value-added chemicals in atmospheric pressure dielectric barrier discharges. Plasma Sources Science and Technology. 2010;**19**:034015

[6] Yu Q, Kong M, Liu T, Fei J, Zheng X. Characteristics of the decomposition of CO_2 in a dielectric packed-bed plasma reactor. Plasma Chemistry and Plasma Processing. 2012;**32**:153-163

[7] Mei D, Zhu X, Wu C, Ashford B, Williams PT, Tu X. Plasma-photocatalytic conversion of CO_2 at low temperatures: Understanding the synergistic effect of plasma-catalysis. Applied Catalysis B: Environmental. 2016;**182**:525-532

[8] Duan X, Li Y, Ge W, Wang B. Degradation of CO_2 through dielectric barrier discharge microplasma. Greenhouse Gases: Science and Technology. 2015; **5**:131-140

[9] Mei D, He YL, Liu S, Yan J, Tu X. Optimization of CO_2 conversion in a cylindrical dielectric barrier discharge reactor using design of experiments. Plasma Processes and Polymers. 2016;**13**:544-556

[10] Mahammadunnisa S, Reddy L, Ray D, Subrahmanyam C, Whitehead JC. CO_2 reduction to syngas and carbon nanofibres by plasma-assisted in situ decomposition of water. International Journal of Greenhouse Gas Control. 2013;**16**:361-363

[11] Aerts R, Somers W, Bogaerts A. Carbon dioxide splitting in a dielectric barrier discharge plasma: A combined experimental and computational study. ChemSusChem. 2015;**8**:702-716

[12] Ozkan A, Dufour T, Silva T, Britun N, Snyders R, Bogaerts A, et al. The influence of power and frequency on the filamentary behavior of a flowing DBD—application to the splitting of CO_2. Plasma Sources Science and Technology. 2016;**25**:025013

[13] Snoeckx R, Heijkers S, Van Wesenbeeck K, Lenaerts S, Bogaerts A. CO_2 conversion in a dielectric barrier discharge plasma: N_2 in the mix as a helping hand or problematic impurity? Energy and Environmental Science. 2016;**9**:999-1011

[14] Duan XF, Hu ZY, Li YP, Wang BW. Effect of dielectric packing materials on the decomposition of carbon dioxide using DBD microplasma reactor. AIchE Journal. 2015;**61**:898-903

[15] Mei D, Tu X. Conversion of CO_2 in a cylindrical dielectric barrier discharge reactor: Effects of plasma processing parameters and reactor design. Journal of CO_2 Utilization. 2017;**19**:68-78

[16] Van Laer K, Bogaerts A. Improving the conversion and energy efficiency

of carbon dioxide splitting in a zirconiaðpacked dielectric barrier discharge reactor. Energy Technology. 2015;**3**:1038-1044

[17] Zhang K, Zhang G, Liu X, Phan AN, Luo K. A Study on CO_2 Decomposition to CO and O_2 by the combination of catalysis and dielectric-barrier discharges at low temperatures and ambient pressure. Industrial and Engineering Chemistry Research. 2017;**56**:3204-3216

[18] Ray D, Subrahmanyam C. CO_2 decomposition in a packed DBD plasma reactor: Influence of packing materials. RSC Advances. 2016;**6**:39492-39499

[19] Brock SL, Marquez M, Suib SL, Hayashi Y, Matsumoto H. Plasma decomposition of CO_2 in the presence of metal catalysts. Journal of Catalysis. 1998;**180**:225-233

[20] Spencer LF, Gallimore AD. Efficiency of CO_2 dissociation in a radio-frequency discharge. Plasma Chemistry and Plasma Processing. 2011;**31**:79-89

[21] Spencer LF, Gallimore AD. CO_2 dissociation in an atmospheric pressure plasma/catalyst system: A study of efficiency. Plasma Sources Science and Techenology. 2013;**22**:015019

[22] Asisov RI, Givotov VK, Krasheninnikov EG, Potapkin BV, Rusanov VD, Fridman A. Soviet Physics–Doklady. 1983;**271**:94

[23] Chen G, Georgieva V, Godfroid T, Snyders R, Delplancke-Ogletree M-P. Plasma assisted catalytic decomposition of CO_2. Applied Catalysis B: Environmental. 2016;**190**:115-124

[24] Silva T, Britun N, Godfroid T, Snyders R. Optical characterization of a microwave pulsed discharge used for dissociation of CO_2. Plasma

Sources Science and Technology. 2014;**23**:025009

[25] Chen G, Britun N, Godfroid T, Georgieva V, Snyders R, Delplancke-Ogletree M-P. An overview of CO_2 conversion in a microwave discharge: the role of plasma-catalysis. Journal of Physics D: Applied Physics. 2017;**50**: 084001

[26] van Rooij GJ, van den Bekerom DCM, den Harder N, Minea T, Berden G, Bongers WA, et al. Taming microwave plasma to beat thermodynamics in CO_2 dissociation. Faraday Discussions. 2015;**183**:233

[27] Chen G, Silva T, Georgieva V, Godfroid T, Britun N, Snyders R, et al. Simultaneous dissociation of CO_2 and H_2O to syngas in a surface-wave microwave discharge. International Journal of Hydrogen Energy. 2015;**40**:3789-3796

[28] Chen G, Godfroid T, Georgieva V, Britun N, Delplancke-Ogletree M-P, Snyders R. Plasma-catalytic conversion of CO_2 and CO_2/H_2O in a surface-wave sustained microwave discharge. Applied Catalysis B: Environmental. 2017;**214**: 114-125

[29] Britun N, Silva T, Chen G, Godfroid T, Mullen J, Snyders R. Plasma-assisted CO_2 conversion: Optimizing performance via microwave power modulation. Journal of Physics D: Applied Physics. 2018;**51**:144002

[30] Mikoviny T, Kocan M, Matejcik S, Mason NJ, Skalny JD. Experimental study of negative corona discharge in pure carbon dioxide and its mixtures with oxygen. Journal of Physics D: Applied Physics. 2004;**37**:64

[31] Nunnally T, Gutsol K, Rabinovich A, Fridman A, Gutsol A, Kemoun A. Dissociation of CO_2 in a low current gliding arc plasmatron. Journal of

Physics D: Applied Physics. 2011;**44**: 274009

[32] Indarto A, Yang DR, Choi JW, Lee H, Song HK. Gliding arc plasma processing of CO_2 conversion. Journal of Hazardous Materials. 2007;**146**: 309-315

[33] Sun SR, Wang HX, Mei DH, Tu X, Bogaerts A. CO_2 conversion in a gliding arc plasma: Performance improvement based on chemical reaction modeling. Journal of CO_2 Utilization. 2017;**17**: 220-234

[34] Neyts EC, Ostrikov K, Sunkara MK, Bogaerts A. Plasma catalysis: Synergistic effects at the nanoscale. Chemical Reviews. 2015;**115**:13408-13446

[35] Chen G, Britun N, Godfroid T, Delplancke-Ogletree MP, Snyders R. Role of Plasma Catalysis in the Microwave Plasma-Assisted Conversion of CO_2. Green Processing and Synthesis. Rijeka, Croatia: Intech Publishing. 2017. ISBN: 978-953-51-5330-6

[36] Fridman A, Rusanov VD. Theoretical basis of non-equilibrium near atmospheric pressure plasma chemistry. Pure and Applied Chemistry. 1994;**66**:1267-1278

[37] Bogaerts A, Kozàk T, Van Laer K, Snoeckx R. Plasma-based conversion of CO_2: Current status and future challenges. Faraday Discussions. 2015;**183**:217-232

[38] Britun N, Chen G, Silva T, Godfroid T, Delplancke-Ogletree M-P, Snyders R. Green Processing and Synthesis. Rijeka, Croatia: Intech Publishing. 2017. ISBN: 978-953-51-5330-6

[39] Wen Y, Jiang X. Decomposition of CO_2 using pulsed corona discharges combined with catalyst. Plasma Chemistry and Plasma Processing. 2001;**21**:665-678

[40] Wang S, Zhang Y, Liu X, Wang X. Enhancement of CO_2 conversion rate and conversion efficiency by homogeneous discharges. Plasma Chemistry and Plasma Processing. 2012;**32**(5):979-989

[41] Li R, Tang Q, Yin S, Sato T. Investigation of dielectric barrier discharge dependence on permittivity of barrier materials. Applied Physics Letters. 2007;**90**:131502

[42] Li R, Tang Q, Yin S, Sato T. Plasma catalysis for CO_2 decomposition by using different dielectric materials. Fuel Processing Technology. 2006;**87**: 617-622

[43] Indarto A, Choi JW, Lee H, Song HK. Conversion of CO_2 by gliding arc plasma. Environmental Engineering Science. 2006;**23**:1033-1043

[44] Silva T, Britun N, Godfroid T, Snyders R. Understanding CO_2 decomposition in microwave plasma by means of optical diagnostics. Plasma Processes and Polymers. 2017;**14**: 1600103

[45] Belov I, Paulussen S, Bogaerts A. Appearance of a conductive carbonaceous coating in a CO_2 dielectric barrier discharge and its influence on the electrical properties and the conversion efficiency. Plasma Sources Science and Technology. 2016;**25**:015023

[46] Ashford B, Tu X. Non-thermal plasma technology for the conversion of CO_2. Current Opinion in Green and Sustainable Chemistry. 2017;**3**:45-49

[47] Xu S, Whitehead JC, Martin PA. CO_2 dissociation in a packed bed DBD reactor: First steps towards a better understanding of plasma catalysis. Chemical Engineering Journal. 2017;**326**:477-488

[48] Zou J, Liu C. Utilization of Carbon Dioxide through Nonthermal

Plasma Approaches. Weinheim: Wiley-VCH Press; 2010. ISBN: 978-3-527-32475-0

[49] Ray D, Saha R, Ch S. DBD plasma assisted CO_2 Decomposition: Influence of diluent gases. Catalysts. 2017;7:244

[50] Snoeckx R, Ozkan A, Reniers F, Bogaerts A. The quest for value-added products from carbon dioxide and water in a dielectric barrier discharge: A chemical kinetics Study. ChemSusChem. 2017;**10**:409-424

[51] Heijkers S, Snoeckx R, Kozák T, Silva T, Godfroid T, Britun N, et al. CO_2 conversion in a microwave plasma reactor in the presence of N2: Elucidating the role of vibrational levels. Journal of Physical Chemistry C. 2017;**119**(23):12815-12828

[52] Wang T, Liu H, Xiong X, Feng X. Conversion of carbon dioxide to carbon monoxide by pulse dielectric barrier discharge plasma. IOP Conference Series: Earth and Environmental Science. 2017;**52**:012100

[53] Fridman A, Kennedy LA. Plasma Physics and Engineering. New York: Taylor and Francis; 2011

[54] Spencer LF. The study of CO_2 conversion in a microwave plasma/catalyst system [PhD dissertation]. Michigan: The University of Michigan; 2012

[55] Whitehead JC. Plasma–catalysis: The known knowns, the known unknowns and the unknown unknowns. Journal of Physics D: Applied Physics. 2016;**49**:243001

[56] Chen HL, Lee HM, Chen SH, Chao Y, Chang MB. Review of plasma catalysis on hydrocarbon reforming for hydrogen production—interaction, integration, and prospects. Applied Catalysis, B: Environmental. 2008;**85**:1-9

[57] Liu C, Xu G, Wang T. Non-thermal plasma approaches in CO_2 utilization. Fuel Processing Technology. 1999;**58**:119-134

[58] Tu X, Whitehead JC. Plasma-catalytic dry reforming of methane in an atmospheric dielectric barrier discharge: Understanding the synergistic effect at low temperature. Applied Catalysis, B: Environmental. 2012;**125**:439-448

[59] Whitehead JC. Plasma catalysis: A solution for environmental problems. Pure and Applied Chemistry. 2010;**82**:1329-1336

[60] Kim HH, Teramoto Y, Ogata A, Takagi H, Nanba T. Plasma catalysis for environmental treatment and energy applications. Plasma Chemistry and Plasma Processing. 2016;**36**:45-72

[61] Neyts EC. Plasma-surface interactions in plasma catalysis. Plasma Chemistry and Plasma Processing. 2016;**36**:185-212

[62] Wang Q, Yan BH, Jin Y, Cheng Y. Dry reforming of methane in a dielectric barrier discharge reactor with Ni/Al$_2$O$_3$ catalyst: Interaction of Catalyst and Plasma. Energy Fuels. 2009;**23**:4196-4201

[63] Zhang AJ, Zhu AM, Guo J, Xu Y, Shi C. Conversion of greenhouse gases into syngas via combined effects of discharge activation and catalysis. Chemical Engineering Journal. 2010;**156**:601-606

[64] Siemens W. Ueber die elektrostatische induction und die verzögerung des stroms in flaschendrähten. Annalen der Physik und Chemie. 1857;**178**:66-122

[65] Lebedev YA. Microwave discharges: generation and diagnostics. Journal of Physics: Conference Series. 2010;**257**:012016

[66] Aerts R. Experimental and computational study of dielectric barrier discharges for environmental applications [PhD thesis]. Antwerpen: Universiteit Antwerpen; 2014

[67] Lee J, Sorescu DC, Deng X. Electron-induced dissociation of CO_2 on TiO_2 (110). Journal of the American Chemical Society. 2011;**133**:10066-10069

[68] Liu L, Li Y. Understanding the Reaction Mechanism of Photocatalytic Reduction of CO_2 with H_2O on TiO_2-Based Photocatalysts: A Review. Aerosol and Air Quality Research. 2014;**14**: 453-469